AF800846

Human-Environment Interactions

Volume 11

Series Editor

Emilio F. Moran, Michigan State University, Bloomington, IN, USA

The *Human-Environment Interactions* series invites contributions addressing the role of human interactions in the earth system. It welcomes titles on sustainability, climate change and societal impacts, global environmental change, tropical deforestation, reciprocal interactions of population-environment-consumption, large-scale monitoring of changes in vegetation, reconstructions of human interactions at local and regional scales, ecosystem processes, ecosystem services, land use and land cover change, sustainability science, environmental policy, among others. The series publishes authored and edited volumes, as well as textbooks. It is intended for environmentalists, anthropologists, historical, cultural and political ecologists, political geographers, and land change scientists.

Human-environment interaction provides a framework that brings together scholarship sharing both disciplinary depth and interdisciplinary scope to examine past, present, and future social and environmental change in different parts of the world. The topic is very relevant since human activities (e.g. the burn of fossil fuels, fishing, agricultural activities, among others) are so pervasive that they are capable of altering the earth system in ways that could change the viability of the very processes upon which human and non-human species depend.

Bhanwar Vishvendra Raj Singh · Amit Kumar Batar
Editors

Sustainable Local Development for Environmental and Social Sustainability

Springer

Editors
Bhanwar Vishvendra Raj Singh
Department of Geography, Faculty of Earth Sciences
Mohanlal Sukhadia University
Udaipur, Rajasthan, India

Amit Kumar Batar
Center for Climate Change Adaptation
National Institute for Environmental Studies
Tsukuba, Japan

ISSN 2214-2339 ISSN 2452-1744 (electronic)
Human-Environment Interactions
ISBN 978-3-031-67302-3 ISBN 978-3-031-67303-0 (eBook)
https://doi.org/10.1007/978-3-031-67303-0

© The Editor(s) (if applicable) and The Author(s), under exclusive license to Springer Nature Switzerland AG 2024

This work is subject to copyright. All rights are solely and exclusively licensed by the Publisher, whether the whole or part of the material is concerned, specifically the rights of translation, reprinting, reuse of illustrations, recitation, broadcasting, reproduction on microfilms or in any other physical way, and transmission or information storage and retrieval, electronic adaptation, computer software, or by similar or dissimilar methodology now known or hereafter developed.
The use of general descriptive names, registered names, trademarks, service marks, etc. in this publication does not imply, even in the absence of a specific statement, that such names are exempt from the relevant protective laws and regulations and therefore free for general use.
The publisher, the authors and the editors are safe to assume that the advice and information in this book are believed to be true and accurate at the date of publication. Neither the publisher nor the authors or the editors give a warranty, expressed or implied, with respect to the material contained herein or for any errors or omissions that may have been made. The publisher remains neutral with regard to jurisdictional claims in published maps and institutional affiliations.

This Springer imprint is published by the registered company Springer Nature Switzerland AG
The registered company address is: Gewerbestrasse 11, 6330 Cham, Switzerland

If disposing of this product, please recycle the paper.

Preface

In our modern era, the paramount question that humanity grapples with is achieving harmony with the natural world. The challenges of our times encompass not only socio-economic complexities but also pressing environmental concerns. Central to this discourse is the imperative to attain Environmental and Social Sustainability (ESS) while safeguarding the delicate balance of our ecosystems, even as our finite natural resources dwindle.

This book, *Sustainable Local Development for Environmental and Social Sustainability*, seeks to shed light on the multifaceted dimensions of sustainability, with a particular focus on local communities and their interactions with the environment. The journey begins with Chap. 1, "Sustainable Local Development: A Pathway to Social and Environmental Sustainability," which serves as the cornerstone of our exploration. As we navigate the challenges posed by climate change, land use changes, and shifting urban landscapes, it becomes evident that sustainable development is not merely an aspiration but an urgent necessity.

This collection of chapters draws upon diverse research endeavors and case studies, each offering unique insights into the complex interplay between human activities, ecological systems, and the quest for sustainable livelihoods. Geoscience methods range from traditional field techniques and conventional data collection to the use of remote sensing and geographical information systems, as well as computer-aided techniques to advance geostatistical and dynamic modeling.

From the pivotal role of fuelwood in the livelihood security of tribal communities in Himachal Pradesh to the investigation of forest degradation in a vital tiger corridor in Rajasthan, each chapter deepens our understanding of the challenges and opportunities for sustainable development. Further chapters transport us to the southern coastal region of Gujarat, where land use and land cover changes are scrutinized for sustainable coastal management. We navigate the intricate dynamics of tidal range and dam-induced flooding in the Narmada Estuarine zone, emphasizing the importance of harmonizing nature's rhythms with human livelihoods.

The urban landscape takes center stage as we shift our attention to East Kolkata, unraveling the changing discourse of urbanism and the significance of wetlands. We explore the Panchet Reservoir's catchment area, detecting land use and land cover changes, which are intricately tied to sustainable development and management. Tourism's role in sustainable cities comes under scrutiny as we investigate tourist satisfaction in Udaipur City, India, and its implications for urban sustainability. The journey culminates in the Veniar River Basin, where Geographic Information System (GIS) technologies are harnessed to delineate suitable land for agricultural sustainability.

The book consists of 11 chapters that contribute to a better recognition and understanding of rapid development in the context of regional changes. This book integrates attributes based broadly on biophysical and human dimensions in spatiotemporal perspectives toward social and environmental susceptibility in India. We hope that this book will deepen the awareness of numerous changes in both natural and human systems that significantly influence the present and future well-being of Indian societies in general. We believe that the book will also be beneficial to people in other fields such as geology, ecology, environmental sciences, tourism, disaster management, and urban studies.

We appreciate the hard work of the contributors and referees for their time and prompt responses. We are especially thankful to series editor Emilio F. Moran, publishing editor Andrea Brody-Barre and Ranjan Muthu for providing assistance. Our deepest thanks go to senior editor Juliana Pitanguy from Springer, Netherlands, for her valuable support throughout the publication process of this book.

Udaipur, India — Bhanwar Vishvendra Raj Singh
Tokyo, Japan — Amit Kumar Batar

Contents

Abbreviations

CBD	Convention on Biological Diversity
CVI	Climate Variable Index
DRR	Disaster Risk Reduction
EKW	East Kolkata Wetlands
ESS	Environmental and Social Sustainability
FCCs	False-Color Composites
GEA	Gender Equality Approach
GEF	The Global Environment Facility
GIS	Geographical Information System
HAD	Human Development Approach
ICLEI	The International Council for Local Environmental Initiatives
ICZM	Integrated Coastal Zone Management
IUCN	The International Union for Conservation of Nature
LDP	Local Development Program
LEAD	Local Empowerment for Accelerated Development
LST	Land Surface Temperature
LULC	Land Use and Land Cover
LVI	Livelihood Vulnerability Index
MEA	Millennium Ecosystem Assessment
MSS	Multisprectral Scanner System
NDVI	Normalized Difference Vegetation Index
NDWI	Normalized Differentiated Water Index
NIR	Near-Infrared
NTFPs	Non-Timber Forest Products
OLI	Operational Land Imager
RDVI	Renormalized Difference Vegetation Index
RTR	Ranthambhore Tiger Reserve
SAVI	Soil-Adjusted Vegetation Index
SDGs	Sustainable Development Goals
SDSS	Spatial Decision Support System
SLD	Sustainable Local Development

SSD	Sardar Sarovar Dam
SWIR	Shortwave Infrared
TIRS	Thermal Infrared Sensor
TLVI	Tribal Livelihood Vulnerability Index
TM	Thematic Mapper
UCCN	The UNESCO Creative Cities Network
UNCED	United Nations Conference on Environment and Development
UNDP	The United Nations Development Program
UNDP	United Nations Development Programme
UNFCCC	The United Nations Framework Convention on Climate Change
USGS	United States Geological Survey

Chapter 1
Sustainable Local Development: A Pathway to Social and Environmental Sustainability

Amit Kumar Batar, Bhanwar Vishvendra Raj Singh, Manjit Singh, and Arun Pratap Mishra

Abstract The essential question that contemporary society must address is how to live-in harmony with nature. Modern society faces numerous socio-economic challenges and environmental issues together—the concern still needs to be addressed of how to bring environmental and social sustainability (ESS) without harming nature, despite limited natural resources. Therefore, in the last few years, the concept of sustainable local development (SLD) has emerged worldwide to tackle localized socio-economic challenges and environmental issues in a way that is sustainable for future generations. This chapter explores the concept of sustainable local development as an approach to addressing socio-economic challenges and environmental issues. This chapter explains how sustainable local development can be linked to global agendas and programs such as the Sustainable Development Goals (SDGs), the Paris Agreement, Convention on Biological Diversity (CBD), and Disaster Risk Reduction (DRR). It also highlights the importance of involving local communities and organizations in the planning and implementation of sustainable development initiatives program such as the United Nations Development Program (UNDP), the

A. K. Batar (✉)
National Institute for Environmental Studies, Tsukuba, Ibaraki, Japan
e-mail: batar.amitkumar@nies.go.jp

B. V. R. Singh
Department of Geography, Faculty of Earth Sciences, Mohanlal Sukhadia University, Udaipur, India
e-mail: bhanwarsa28@mlsu.ac.in

M. Singh
Department of Geography, School of Earth, Biological and Environment Sciences, Central University of South Bihar, Gaya, Bihar, India
e-mail: manjitsingh@cusb.ac.in

A. P. Mishra
Department of Geography, Delhi School of Economics, University of Delhi, New Delhi, India
e-mail: apmishradse@gmail.com

© The Author(s), under exclusive license to Springer Nature Switzerland AG 2024
B. V. Raj Singh and A. K. Batar (eds.), *Sustainable Local Development for Environmental and Social Sustainability*, Human-Environment Interactions 11,
https://doi.org/10.1007/978-3-031-67303-0_1

International Union for Conservation of Nature (IUCN), and the Global Environment Facility (GEF). The chapter identifies common challenges faced in implementing sustainable local development, such as lack of coordinated and integrated planning, lack of capacity and resources, and lack of participation and engagement of local communities and stakeholders. It also examines strategies for overcoming these challenges, such as increasing funding, implementing policies and regulations, promoting public education and engagement, involving diverse stakeholders, and aligning sustainable development initiatives with global agreements and conventions.

Keywords Sustainable local development · Social and environmental sustainability · Local communities · Natural resources · Biodiversity conservation · Climate change mitigation · Global agendas

1.1 Introduction

The essential question that contemporary society must address is how to live-in harmony with nature. Modern society faces numerous socio-economic challenges and environmental issues together—the concern still needs to be addressed regarding how to bring environmental and social sustainability (ESS) without harming nature in limited natural resources. Although technology development can help access electricity, food, health, education, and many other facilities, the risks of climate change and depletion of natural resources negatively impact livelihood security, biodiversity loss, and deterioration of ecosystem services. These issues emerge due to anthropogenic pressures and encroachments into the natural environment. As a result, the impact of these issues is already visible worldwide based on different socio-economic, political, geographical, and cultural factors. Consequently, these issues are significant concerns for scientists, stakeholders, and policymakers on how to bring environmental and social sustainability (ESS) in changing environments without harming nature. Specifically, infrastructure development and natural resource conservation needs may help socio-economic recovery and environment conservation in the short term but cannot be considered sustainable development when considering future generations. Consequently, in the last few years, the concept of sustainable local development (SLD) has emerged worldwide to tackle localized socio-economic challenges and environmental issues in a way that is sustainable for future generations. The objective of this chapter is to examine the concept of sustainable local development and its potential for addressing socio-economic challenges and environmental issues, while contributing to the global Sustainable Development Goals and climate change mitigation. Additionally, this chapter will examine the challenges and solutions for implementing sustainable local development.

1.2 Sustainable Local Development: An Overview of Its Evolution and Definition

Sustainable local development is a concept that has evolved over time and has been defined in different ways by various organizations and experts. Sustainable local development (SLD) integrates the idea of sustainability, which was first introduced in the 1970s, into local development (Amin, 1999). One early reference to sustainable local development can be found in the 1987 Brundtland Report, also known as "Our Common Future," which was published by the United Nations World Commission on Environment and Development. The report defined sustainable development as "development that meets the needs of the present without compromising the ability of future generations to meet their own needs". The report also emphasized the importance of involving local communities in the planning and implementation of sustainable development initiatives and stressed the need to protect the natural environment and promote social equity. Another early reference to sustainable local development can be found in the 1992 United Nations Conference on Environment and Development (UNCED), also known as the Earth Summit, which was held in Rio de Janeiro, Brazil. The conference adopted Agenda 21, a global action plan for sustainable development, which included a chapter on sustainable local development. The chapter emphasized the importance of involving local communities in decision-making and planning processes and called for the development of local capacity and partnerships to support sustainable development initiatives. According to the United Nations Development Program (UNDP), sustainable local development is "an approach that aims to improve the quality of life for all citizens, taking into account the economic, social and environmental dimensions of development, and that involves and empowers local communities in the planning, implementation, monitoring, and evaluation of development initiatives" (United Nations Development Programme, 2017). The European Union defines sustainable local development as "a process of economic, social and environmental regeneration, which aims to improve the quality of life for all citizens, and to ensure that the benefits of development are shared equitably, both within and between generations (European Commission, 2018). In recent years, the goal of sustainable local development has been to improve the standard of living and well-being of the current population while also considering the impact on future generations. This approach emphasizes the importance of maintaining and preserving the natural environment and respecting its constraints. In 2016, the United Nations Development Program (UNDP) launched the Sustainable Development Goals strategy as a continuation of the Millennium Development Goals, highlighting the significance of SLD in today's world (United Nations, 2016). Local development, in general, is a process of improving the economic, social, and environmental situation of a given area by utilizing the resources that are available within that area. This approach is characterized by using endogenous resources, which take advantage of the economic and social dynamics of a specific territory and are crucial for the past and future development of the area (Milán-García et al., 2019). SLD focuses on achieving a balance between economic, social, and environmental objectives without compromising natural resources. Thus, SLD refers to transformations that can be carried out in a territory once the economic, social, and environmental dimensions have been balanced. SLD must have the capacity to provide feedback and maintain itself over time, respecting the constraints presented by the surrounding natural environment (Mebratu, 1998).

1.3 Social and Environmental Sustainability in Local Development: Why It Matters

Social and environmental sustainability are critical components of local development, as they ensure the well-being of current and future generations is protected. Addressing social sustainability in local development includes addressing issues such as poverty, inequality, and access to basic services, which are critical for achieving the Sustainable Development Goals (SDGs). Environmental sustainability, on the other hand, includes protecting natural resources, reducing greenhouse gas emissions, and promoting conservation and sustainable use of biodiversity. Integrating social and environmental sustainability into local development ensures that development is inclusive and equitable and respects the rights of all members of the community, including marginalized and vulnerable groups. Furthermore, addressing social and environmental sustainability in local development can help to build resilience to climate change, disasters, and other environmental risks, leading to more sustainable and livable communities. Failure to address these issues in local development can lead to increased poverty and inequality, loss of biodiversity, and increased vulnerability to climate change. Therefore, addressing social and environmental sustainability in local development is essential for the well-being of current and future generations.

1.4 Linking Sustainable Local Development with Global Agendas and Programs

The sustainable local development approach can help international agendas and programs by providing a framework and guidance for local communities and organizations to develop and implement initiatives that improve their quality of life in a socially and environmentally sustainable way. By involving local communities and organizations in the planning and implementation of sustainable development initiatives, international agendas and programs can better address the needs and challenges of local communities and support their efforts to achieve sustainable development.

1.4.1 SDGs and Sustainable Local Development

The Sustainable Development Goals (SDGs) are a set of seventeen goals and one hundred and sixty-nine targets adopted by the United Nations in 2015 as a universal call to action to end poverty, protect the planet, and ensure that all people enjoy peace and prosperity by 2030. A sustainable local development approach can help in achieving the SDGs by addressing issues such as poverty, unemployment, and access to basic services, promoting sustainable consumption and production patterns and responsible management of natural resources and ecosystems, reducing urban–rural disparities, promoting sustainable agriculture and rural development, supporting the development of sustainable cities and communities, and building adaptation and resilience by building the capacity of local communities and organizations to prepare for and respond to the impacts of disasters. By implementing sustainable local development strategies and programs, it can contribute to achieving the global Sustainable Development Goals outlined in the United Nations' 2030 Agenda for Sustainable Development.[1,2,3,4,5]

1.4.2 The Paris Agreement and Sustainable Local Development

The Paris Agreement is an international agreement adopted by the United Nations Framework Convention on Climate Change (UNFCCC) in 2015. It aims to combat climate change and reduce greenhouse gas emissions, with the goal of keeping the global temperature rise this century well below 2 °C above pre-industrial levels and to pursue efforts to limit the temperature increase even further to 1.5 °C. A sustainable local development approach can help in the implementation of the Paris Agreement by promoting decentralized climate action, which is critical for achieving the goals of the Paris Agreement, by involving local communities and organizations in the planning and implementation of climate change mitigation and adaptation measures. This approach can help to achieve a low-carbon society by promoting sustainable consumption and production patterns, and responsible management of natural resources and ecosystems. Additionally, it can help in building adaptation and resilience by building the capacity of local communities and organizations to prepare for and respond to the impacts of climate change and fostering sustainable

[1] "Localizing the Sustainable Development Goals" by United Nations Development Programme (UNDP).

[2] "Sustainable Local Development: An Overview" by United Nations Development Programme (UNDP).

[3] "Sustainable Development Goals" by United Nations.

[4] "2030 Agenda for Sustainable Development" by United Nations.

[5] "Sustainability in Local Development: A Guide for Practitioners" by International Labour Organization (ILO).

urban and rural development, which is crucial to reducing greenhouse gas emissions and increasing the resilience of communities to the impacts of climate change. Furthermore, it can help to ensure that the Paris Agreement is being implemented at the local level and that local priorities are being considered.[6,7,8,9,10,11,12]

1.4.3 Convention on Biological Diversity (CBD) and Sustainable Local Development

The history of biodiversity framework began in 1972 with the recognition of the need to conserve biodiversity as a global issue, leading to the creation of the Convention on Biological Diversity (CBD) in 1992. The CBD aims to conserve biodiversity, sustainably use its components, and share the benefits of genetic resources fairly and equitably. In 2002, the CBD adopted the Strategic Plan for Biodiversity 2011–2020 and its Aichi Biodiversity Targets, which set out specific targets for conservation, sustainable use, and benefit sharing by 2020. In 2020, the Convention on Biological Diversity adopted the post-2020 Global Biodiversity Framework, a new strategic plan for the conservation and sustainable use of biodiversity for the period 2021–2030. This framework aims to conserve biodiversity, protect, and restore ecosystems, and ensure fair and equitable sharing of benefits from the use of genetic resources. A sustainable local development approach can help in achieving the objectives of the CBD by promoting decentralized action and involving local communities and organizations in the planning and implementation of biodiversity conservation and sustainable use measures. It also supports sustainable economic growth by promoting sustainable consumption and production patterns, and responsible management of natural resources and ecosystems, which can conserve biodiversity. Additionally, it promotes adaptation and resilience by building the capacity of local communities and organizations to prepare for and respond to the impacts of disasters and climate

---

[6] See Footnote 2.

[7] Decentralized Climate Action" by United Nations Framework Convention on Climate Change (UNFCCC).

[8] Adaptation and Resilience" by United Nations Framework Convention on Climate Change (UNFCCC).

[9] The Paris Agreement" by United Nations Framework Convention on Climate Change (UNFCCC).

[10] The Low-Carbon Society Vision in Japan: A Path to Sustainable Development (http://jst.go.jp/global/english/policy/low-carbon/index.html).

[11] Low-carbon Development Strategies in Cities (http://iclei.org/fileadmin/user_upload/documents/LCDS_in_Cities_Report.pdf).

[12] Low-Carbon Cities: Achieving a Low-Carbon Society through City-Based Climate Action (http://c40.org/research/low-carbon-cities).

change, which are threats to biodiversity. By implementing sustainable local development strategies and programs, it can contribute to achieving the global biodiversity goals outlined in the Convention on Biological Diversity (CBD).[13,14,15,16,17]

1.4.4 Disaster Risk Reduction and Sustainable Local Development

Disaster Risk Reduction (DRR) is the process of identifying, assessing, and reducing the risks of natural and man-made hazards. It aims to reduce the impact of disasters on communities and economies and to build resilience to future hazards. A sustainable local development approach can help in DRR by incorporating measures into local development plans and programs. This approach promotes inclusive and participatory development, which involves local communities and stakeholders in the identification, assessment, and reduction of disaster risks. It also supports sustainable economic growth by promoting sustainable consumption and production patterns, and responsible management of natural resources and ecosystems, which can reduce the impact of natural hazards on local communities. Additionally, it helps in building adaptation and resilience by building the capacity of local communities and organizations to prepare for and respond to the impacts of disasters. Furthermore, it fosters sustainable urban and rural development, which is crucial to reducing disaster risks and increasing the resilience of communities to hazards.[18,19,20]

---

[13] Convention on Biological Diversity (CBD)—Secretariat of the Convention on Biological Diversity (http://cbd.int/convention/).

[14] CBD—United Nations Framework Convention on Climate Change (http://unfccc.int/bodies/body/convention-on-biological-diversity).

[15] The Role of Local Governments in Implementing the Convention on Biological Diversity (http://cbd.int/localgov/).

[16] Local Governments and Biodiversity Conservation (http://iucn.org/resources/issues-briefs/local-governments-and-biodiversity-conservation).

[17] Sustainable Consumption and Production (http://un.org/sustainabledevelopment/sustainable-consumption-production).

[18] Disaster Risk Reduction and Sustainable Development (http://unisdr.org/we/inform/publications/43291).

[19] Guidelines for the Integration of Disaster Risk Reduction in Sustainable Development (http://preventionweb.net/publications/view/49138).

[20] Sustainable Development and Disaster Risk Reduction (http://preventionweb.net/publications/view/49364).

1.4.5 The United Nations Development Program (UNDP) and Sustainable Local Development

The United Nations Development Program (UNDP) is a leading global agency that provides technical and financial support to countries to help them achieve the Sustainable Development Goals (SDGs). It uses a sustainable local development approach to support local communities and governments in the planning and implementation of sustainable development initiatives. This approach helps address local issues such as poverty, unemployment, and access to basic services, build local capacity, promote inclusive and participatory development, support sustainable economic growth, foster sustainable urban and rural development, and localize the SDGs. It should be noted that the UNDP also uses other approaches, such as the Human Development Approach (HDA) and Gender Equality Approach (GEA), in addition to the sustainable local development approach. One of the key examples of UNDP's work in sustainable local development is the Local Development Program (LDP), which aims to improve the economic and social well-being of people living in poverty, particularly in rural areas. Another successful initiative is the Local Empowerment for Accelerated Development (LEAD) program, which is implemented in many countries by UNDP. It focuses on empowering communities to become more active and engaged in local governance and development through building capacities of local government and civil society organizations. The program aims to enhance the participation of local communities in decision-making processes and to improve the effectiveness of local governments in delivering services.[21,22]

1.4.6 The Global Environment Facility (GEF) and Sustainable Local Development

The Global Environment Facility (GEF) is an international financing organization that helps developing countries fund projects and programs related to biodiversity, climate change, international waters, land degradation, and chemicals and waste management. The GEF uses a sustainable local development approach to support initiatives that benefit local communities and the environment. For example, the GEF has supported the development of sustainable agriculture and forestry projects in Nepal, which have helped to protect natural resources and improve the livelihoods of local communities. The GEF has implemented several programs that focus on sustainable local development, such as the Small Grants Programme (SGP), which supports community-based projects that promote sustainable development and biodiversity

[21] "UNDP and Local Development"—United Nations Development Programme: http://undp.org/content/undp/en/home/ourwork/povertyreduction/focus_areas/local_development.html.

[22] "Local Empowerment for Accelerated Development (LEAD)"—United Nations Development Programme: http://undp.org/content/undp/en/home/ourwork/povertyreduction/focus_areas/local_empowerment_for_accelerated_development.html.

conservation. This program provides grants to local NGOs, CBOs, and community-based organizations to implement projects that address environmental issues and promote sustainable livelihoods. Another example is the GEF-7 Sustainable Forest Landscape (SFL) project, which supports countries in adopting sustainable land-use practices. This project aims to promote sustainable forest management, agro-forestry, and sustainable land-use practices, as well as to reduce greenhouse gas emissions from deforestation and forest degradation.[23,24,25,26,27] Thus, a sustainable local development approach can help the GEF in several ways. It can provide a framework for implementing sustainable development projects that consider the economic, social, and environmental aspects of development. This approach can also ensure that the needs and perspectives of local communities are considered, and that they are actively engaged in the planning, implementation, and evaluation of projects.

1.4.7 Habitat III and Sustainable Local Development

Habitat III was a United Nations conference held in Quito, Ecuador, in 2016 to address the challenges of urbanization and create a new global agenda for sustainable urban development. The conference resulted in the adoption of the New Urban Agenda, a comprehensive and action-oriented document that provides a framework for sustainable urban development. The Sustainable Local Development approach, which focuses on the economic, social, and environmental aspects of development and ensures local communities are engaged and considered, aligns well with the goals of the New Urban Agenda. This approach can help to ensure that sustainable urban development is inclusive, holistic, and people-centered by promoting active community engagement, integrating different sectors, and linking urban and rural areas (Habitat III, 2016).

1.4.8 UNESCO's Creative Cities Network and Sustainable Local Development

The UNESCO Creative Cities Network (UCCN) is a program launched by UNESCO in 2004 with the aim of promoting cooperation and sharing best practices among cities

---

[23] Global Environment Facility (GEF)—http://thegef.org/.

[24] GEF's sustainable local development approach—http://thegef.org/about-gef/strategies-approaches/sustainable-local-development.

[25] GEF's support for sustainable agriculture and forestry projects in Nepal—http://thegef.org/project/sustainable-agriculture-and-forestry-project-safp.

[26] GEF's Small Grants Programme (SGP)—http://thegef.org/project/small-grants-programme-sgp.

[27] GEF-7 Sustainable Forest Landscape (SFL) project—http://thegef.org/project/gef-7-sustainable-forest-landscape-sfl.

that have identified creativity as a strategic factor for sustainable urban development. The program aims to enhance cultural diversity, social inclusion, and economic development in the areas of arts, culture, and creative industries. The Sustainable Local Development approach aligns well with the goals of the UCCN, focusing on the economic, social, and environmental aspects of development, and ensuring the active engagement of local communities in the planning, implementation, and evaluation of development projects. This approach can help to create livable, resilient, inclusive, safe, healthy, and sustainable cities by fostering cooperation and sharing best practices among cities and taking into account the interconnections between different sectors and links between urban and rural areas.[28,29]

1.4.9 The International Council for Local Environmental Initiatives (ICLEI) and Sustainable Local Development

The International Council for Local Environmental Initiatives (ICLEI) is a global network of cities, towns, and regions dedicated to promoting sustainable urban development through local government action. ILEI's mission is to build a worldwide movement of local governments to achieve tangible improvements in global sustainability, with a particular focus on local government's role in addressing climate change. The Sustainable Local Development approach can support ICLEI's mission in various ways, by ensuring that the economic, social, and environmental aspects of development are considered and that the needs and perspectives of local communities are taken into account. This aligns with ICLEI's goal of promoting inclusive and sustainable urban development. Additionally, by promoting active engagement of local communities in the planning, implementation, and evaluation of development projects, the approach can help ICLEI to create livable, resilient, inclusive, safe, healthy, and sustainable cities. Furthermore, by fostering cooperation and sharing best practices among cities and considering the interconnections between different sectors and links between urban and rural areas, the approach can help ICLEI to create sustainable and integrated urban–rural development.[30,31,32]

---

[28] UNESCO Creative Cities Network: http://en.unesco.org/creative-cities.

[29] UNESCO Creative Cities Network objectives: http://en.unesco.org/creative-cities/content/objectives.

[30] International Council for Local Environmental Initiatives (ICLEI)—http://iclei.org/.

[31] ICLEI's mission: http://iclei.org/about-iclei/mission/.

[32] ICLEI's approach to sustainable urban development: http://iclei.org/sustainable-urban-development/.

1.4.10 The International Union for Conservation of Nature (IUCN) and Sustainable Local Development

The International Union for Conservation of Nature (IUCN) is a global environmental organization that promotes sustainable development and conservation of natural resources. It works to protect biodiversity and ecosystems, and to ensure the sustainable use of natural resources. The organization's mission is to influence, encourage, and assist societies throughout the world to conserve the integrity and diversity of nature, and to ensure that any use of natural resources is equitable and ecologically sustainable. The Sustainable Local Development approach aligns well with the goals of the IUCN, as it focuses on the economic, social, and environmental aspects of development and ensures that the needs and perspectives of local communities are considered. Additionally, by promoting active engagement of local communities in the planning, implementation, and evaluation of development projects, the approach can help the IUCN to create livable, resilient, inclusive, safe, healthy, and sustainable communities. Furthermore, by fostering cooperation and sharing best practices among cities and considering the interconnections between different sectors and links between urban and rural areas, the approach can help the IUCN to create sustainable and integrated urban–rural development, which is in line with the organization's mission.[33,34]

1.5 Sustainable Local Development: Challenges and Solutions

1.5.1 Challenges of Implementing Sustainable Local Development for Targeting Global Agreements

Sustainable local development is an approach that seeks to improve the well-being of communities while preserving the environment for future generations. It is a holistic approach that considers economic, social, and environmental sustainability in decision-making and planning. This approach can be used to achieve several global goals, including the Sustainable Development Goals (SDGs), the Paris Agreement, the Convention on Biological Diversity (CBD), and Disaster Risk Reduction. However, there are also challenges to implementing sustainable local development. One major challenge is the lack of adequate funding for sustainable projects. Many local governments and organizations lack the resources to invest in sustainable infrastructure and programs. Another challenge is the lack of political will to prioritize sustainable development. Decisionmakers may prioritize short-term

[33] International Union for Conservation of Nature (IUCN)—http://iucn.org/.

[34] IUCN's approach to sustainable development and conservation: http://iucn.org/our-work/sustainability-and-resilience.

economic gains over long-term sustainability. Additionally, there can be a lack of public awareness and participation in sustainable development efforts. Without the support and engagement of the community, sustainable initiatives may struggle to gain momentum.

1.5.2 Challenges of Implementing Sustainable Local Development for a Local Perspective

Sustainable local development is an approach that aims to achieve economic, social, and environmental sustainability at the local level. However, there are also challenges to implementing this approach. One of the main challenges is the lack of coordinated and integrated planning among different sectors and actors. For example, local development plans often do not consider the impacts of climate change and biodiversity loss, and Disaster Risk Reduction (DRR) strategies are often not integrated into local development plans. This lack of integration can lead to conflicting priorities and impede progress in achieving Sustainable Development Goals. Another challenge is the lack of capacity and resources at the local level. Many communities, particularly in developing countries, lack the necessary resources and expertise to implement sustainable development plans and projects. This can make it difficult for local actors to effectively participate in the planning and implementation of sustainable development initiatives. A third challenge is the lack of participation and engagement of local communities and stakeholders in the planning and implementation of sustainable development initiatives. This can lead to a lack of buy-in and ownership of the initiatives, which can make it difficult to achieve Sustainable Development Goals.

1.5.3 Strategies for Successful Sustainable Local Development Implementation

To overcome these challenges, solutions include increasing funding for sustainable projects through government grants and private–public partnerships, implementing policies and regulations that prioritize sustainable development, and promoting public education and engagement in sustainable practices. Additionally, involving a diverse group of stakeholders in the decision-making process, such as community members, can help to ensure that the needs and perspectives of all are considered. It is also important to coordinate and integrate different sectors and actors in the planning and implementation of sustainable development initiatives. This can be done by creating a common vision and shared goals, and by coordinating different sectors and actors in the planning, implementation, and monitoring of sustainable development initiatives.

In addition, it is important to ensure that sustainable development initiatives are aligned with the Sustainable Development Goals (SDGs), the Paris Agreement, the Convention on Biological Diversity (CBD), and Disaster Risk Reduction (DRR). This can be done by incorporating these agreements and conventions into local development plans and by ensuring that sustainable development initiatives are aligned with their goals and targets. Overall, sustainable local development is a complex and challenging process that requires the coordination and integration of different sectors and actors, the involvement of local communities and stakeholders, and the alignment of sustainable development initiatives with the SDGs, the Paris Agreement, the CBD, and DRR. However, by addressing these challenges, it is possible to create resilient and sustainable communities that contribute to the achievement of Sustainable Development Goals.

It is important to recognize that indigenous local knowledge can be a key strategy for implementing Sustainable Local Development and making substantial contributions to global agendas by supporting sustainable practices and preserving cultural and ecological diversity. Indigenous local knowledge plays a crucial role in the successful implementation of sustainable local development and contributes to global agendas. By incorporating the wisdom and practices of indigenous communities, development initiatives can enhance environmental stewardship, promote cultural diversity and inclusivity, foster community participation and ownership, and support sustainable livelihoods. For example, indigenous knowledge about traditional farming practices and natural resource management can contribute to biodiversity conservation, climate change adaptation, and the development of sustainable livelihoods. By valuing and integrating indigenous local knowledge, strategies for sustainable local development can align with global goals and achieve long-term success. Another example is found in a coastal community, where the indigenous local knowledge of traditional fishing practices and marine resource management can make significant contributions to both global agendas and the local perspective of sustainable local development.

1.6 Conclusion

Sustainable local development is a holistic approach that addresses the economic, social, and environmental well-being of a specific community. It aims to achieve long-term, balanced growth while ensuring social and environmental sustainability. The key aspect of sustainable local development is community engagement and participation, which empowers community members to identify their needs and priorities and actively participate in the development process. The integration of social and environmental considerations into decision-making, using sustainable practices and technologies, and the protection and conservation of natural resources, is another important aspect of sustainable local development. The importance of continued efforts to support sustainable local development initiatives emphasizes the need for ongoing support for sustainable development initiatives in local communities, including investing in infrastructure, education, and economic development, as well as promoting environmental sustainability and social inclusion. It also highlights the importance of community engagement and participation in these initiatives to ensure their success, as well as the importance of collaboration among different stakeholders, including governments, businesses, and civil society organizations, to achieve sustainable local development.

References

Amin, A. (1999). An institutionalist perspective on regional economic development. *International Journal of Urban Reginal Research, 23*, 365–378. [Google Scholar] [CrossRef].

European Commission. (2018). Sustainable local development in the Emilia-Romagna region. Retrieved from http://ec.europa.eu/regional_policy/en/policy/themes/sustainable-development/good-practices/sustainable-local-development-in-the-emilia-romagna-region_en

Habitat III, a United Nations conference held in 2016 in Quito, Ecuador. http://unhabitat.org/habitat-iii

Mebratu, D. (1998). Sustainability and sustainable development: Historical and conceptual review. *Environment Impact Asses, 18*, 493–520. [Google Scholar] [CrossRef].

Milán-García, J., Uribe-Toril, J., Ruiz-Real, J. L., & de Pablo Valenciano, J. (2019). Sustainable local development: An overview of the state of knowledge. *Resources, 8*, 31. https://doi.org/10.3390/resources8010031

United Nations. (2016). *Sustainable Development Goals Report.* New York, NY, USA: United Nations. [Google Scholar].

United Nations Development Programme. (2017). Empowering local communities for sustainable development. Retrieved from http://undp.org/content/undp/en/home/sustainable-development-goals/goal-11-sustainable-cities-and-communities/target-11-3-by-2030-enhance-inclusive-and-sustainable-urbanization-and-capacity-for-participatory-integrated-urban-planning-and-design-at-all-levels.html

Chapter 2
Climate Change Livelihood Vulnerability of Tribal Community and Its Determinants: A Cross-sectional Study in Drought-Prone Region of West Bengal, India

Sourav Kumar Das and Aishwarya Basu

Abstract *Introduction*: The tribal livelihoods are mainly dependent on forest resources while they are facing challenges socially and economically due to climate change which makes them vulnerable. According to the 2011 Census of India, 5.7% of West Bengal's total population belongs to Schedule tribes against 8.6% tribal population in India. *Objective*: The paper tries to assess climate change vulnerability by formulating index based on tribal household level. Second, is to find the percentage of households who are more vulnerable tribes in 2015 and in 2020. Third, is to estimate the factors of climate change for which livelihood becomes vulnerability of tribal communities. *Method and Material*: This paper has been based on field survey data collected from fifty (50) households who are tribes in the village Kamo located in Bankura district, a drought-prone district of south West Bengal in the years of 2015 and 2020. The repeated sampling technique was followed for the collection of data. The study has calculated climate change tribal vulnerability index based on modified version of (Hahn et al., Global Environmental Change 19:74–88, 2009) methodology. The calculation of such index consists of the main indicators like index of socio-economy, livelihood, forest dependency, social security and health, and sub-indicators for each main indicator are chosen in consultation with local conditions where the tribal communities live and from the literature review. The model of beta regression is used to determine the factors of livelihood vulnerability of tribes. *Result*: The result of the study showed that the climate change tribal livelihood vulnerability index is found to be 0.49 for 2020, and it is 0.44 for 2015. The mean difference of tribal vulnerability index is statistically significant by "*t*" test. The result further reveals that the percentage of more of tribal vulnerable households (56%) in 2020

S. K. Das (✉)
Department of Economics, Women's College, Calcutta, West Bengal, India
e-mail: souravsouthkolkata@gmail.com

A. Basu
Department of Economics, Burdwan University, West Bengal, Burdwan, India

© The Author(s), under exclusive license to Springer Nature Switzerland AG 2024

B. V. Raj Singh and A. K. Batar (eds.), *Sustainable Local Development for Environmental and Social Sustainability*, Human-Environment Interactions 11,
https://doi.org/10.1007/978-3-031-67303-0_2

is higher than that for the year of 2015 (26%). The climate change tribal livelihood vulnerability is caused by socio-economic variables and climate variable.

Keyword Tribal Livelihood Vulnerability · Climate Change · Forest Dependency · Social Security Programmes

2.1 Introduction

For last few decades, the world is facing major socio-economic and environmental changes due to environmental degradation and climate changes. Change in climate is a global risk to sustainable development because for its impacts on forest ecosystems, health, food security and agriculture (IPCC, 2007). The economy of India is mainly dependent on sectors which are climate sensitive, for example, resources of water, agriculture and zones nearing coastal areas, forestry and biodiversity (INCCA, 2010). About 100 million rural populations of India are directly dependent on those sectors which are climate sensitive. The ecosystems of forest provide social and economic benefits, in the form of food, fodder, employment, as well as cultural values protection (FAO, 2006). Forest-dependent communities become more vulnerable due to the change in climate on forests biodiversity.

Forests are threatened by growing pressure to sustain the fuel and fodder needs due to the increasing human needs and cattle population. In India, about hundred million population based on forests are dependent on non-timber forest products (NTFPs) (Saxena, 2003). The NTFPs play a major role for the millions of urban and rural people over the world (Malhotra & Bhattacharya, 2010; Pandey & Kori, 2011; Shackleton et al., 2015). These NTFPs fulfil various functions for supporting human wellbeing. The natural resources or forests played a role in the economic development locally, and sustainability of forestry has been studied by many researchers. Several researchers (say Adhikari, 2005; Arnold & Bird, 1999; Cavendish, 1999; Reddy & Chakravarty, 1999) studied in developing countries and measured the contribution of forestry and forest products for generating income and also estimated the socio-economic factor which has an impact on forest dependency. Evidence also states that income from non-timber forest products contributing within a range of 10% to 60% of income (Cavendish, 2000; Fisher, 2004; Reddy & Chakravarty, 1999). The contribution of forestry to the tribal livelihood varies due to different nature of forest products, types of forest management and also due to different market values.

According to Census of India (2011), about 5.7% of West Bengal's total population belongs to schedule tribes against 8.6% tribal population in India. Tribal communities residing in an around forest are mainly dependent on forest resources, so they are vulnerable to the impacts of change in climate than other sections of the society (Bhattacharya & Prasad, 2009; Thomas & Twyman, 2005). In most underdeveloped and developing countries, as most of them are poor and underprivileged, they are mainly dependent on forest-based resources such as fuelwood, food, fodder, medicine and timber, from where they are generating money and also fulfilling their needs for subsistence (Angelsen et al., 2014; Mukul et al., 2016). Various studies from developing countries pointed out that forest products play an important role in rural livelihoods, mainly for the tribes. Tribes are vulnerable due to the change in climate due to their close connection and dependence on with the forest resources for their livelihoods, culture and health. So, tribal communities are facing difficult social and economic conditions that may worsen by climatic change and there is a question of their sustainability also.

Hahn et al. (2009) had developed the index of livelihood vulnerability for estimating climate change vulnerability in Mozambique. Ahsan and Warner (2014) developed a socio-economic vulnerability index for climate change-affected livelihoods of Koyra upazila in south-western coastal Bangladesh. Results showed that southern and south-eastern unions are more vulnerable, due to climatic hazards and residing by the Sundarbans mangrove forest. Ghosh and Ghosal (2020) used the index of climate change vulnerability in the flood-prone areas of the foothills of the Himalayan and observed that villages were more vulnerable due to climate change and forest villages are relatively more vulnerable compared to the agricultural villages. Deb and Mukherjee (2022) applied the vulnerability index among the tribes in the Dooars of West Bengal, India, to assess the vulnerability to climate variability. Study shows that due to floods and rainfall, the tribal community becomes vulnerable who are primarily residing in rural region with a least resource base. Das and Basu (2022) measure the climate change livelihood vulnerability across tribes of Lodha, Bhumij, Munda and Santal in West Bengal, India. Results showed that Lodha tribes are more vulnerable than other communities. The reasons of that livelihood vulnerability are the income from agriculture, size of family and poverty.

The above review of literature reveals that there are few researches, particularly quantitative research on the vulnerability of tribal livelihood owing to climatic change over time for schedule tribes of West Bengal, India.

Given this backdrop, this paper tries to measure and compare livelihood vulnerability index of the tribes for the year 2020 with that for 2015. Second, is to calculate the percentages of tribal households who are more vulnerable as well as less vulnerable in 2015 and in 2020 also. Third, is to estimate the factors of climate change vulnerability at household level of tribal communities.

2.2 Materials and Method

This study has been based on field survey data collected from fifty (50) tribal households for the village Kamo located in Bankura district, a drought-prone district of West Bengal in the years of 2015 and 2020. The technique of repeated sampling was followed for collecting data.

2.3 Study Area

Bankura, a backward district in the state of West Bengal, has five tribal-dominated blocks, namely Hirbandh, Ranibandh, Raipur, Chatna and Khatra which are the most deprived parts of the district. Out of these five tribal-dominated blocks, Raipur and Ranibandh blocks come under Jungle-Mahal area of the state West Bengal. Most of the regions of these blocks as well as villages of these blocks are forested as well as drought-prone area. Agriculture and other primary sectors are dependent on rainwater, and paddy is cultivated once in a year if rain permits. The forest, once which was the source of livelihood for tribes, now only providing fuel and generating some income from NTFPs. Though from the collection of Sal leaves, Kendu, flower of Mahua, Neem, Kusum, fruits of Mahua, Haritaki, Amlaki and fruits of Sal, and other kinds of barks and herbs which used as medicine can still generating an alternative livelihood source for the tribes (Fig. 2.1).

2.3.1 Data Collection and Sampling Technique

Ranibandh is constituted by the highest tribal population of Bankura district. Stratified random sampling procedure was followed for the selection of village. Thus, we have selected Kamo village. 50 tribal households from the village Kamo at the

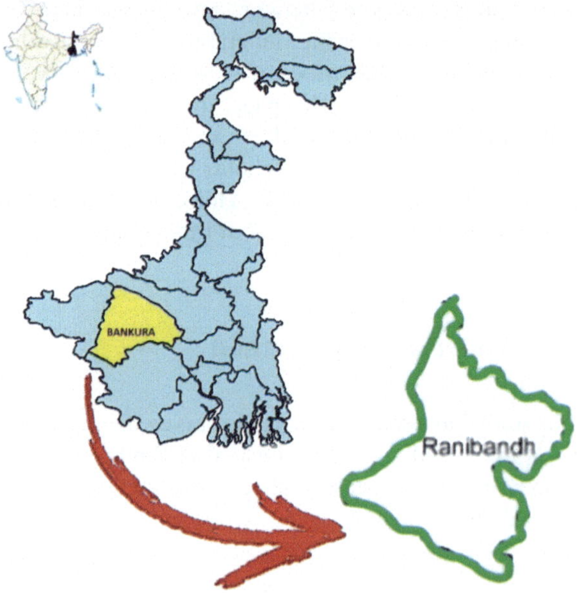

Fig. 2.1 Map of the studied area

Table 2.1 Number of studied households of village Kamo, Bankura district

Bankura	Total households (in numbers)	Sample households (in number)
2015	220	50
2020	226	50

Source Field survey in 2015 and 2020

two point of time say 2015 and 2020 have been collected using repeated sampling techniques. The number of sample households is shown in Table 2.1.

2.3.2 Construction of Tribal Livelihood Vulnerability Index and Climate Variable Index

For measuring the index of livelihood vulnerability, this paper utilizes the index of Livelihood Vulnerability of Hahn et al. (2009) with few modifications. According to Hahn et al. (2009), LVI has seven components: Socio-demographic profile, social networks, livelihood, food, water, and natural disasters and health. Since we are studying tribal livelihood, some new indicators like social security and forest dependency have to be incorporated instead of social network, food and water for the estimation of index of livelihood vulnerability of the tribes. The detailed component of the livelihood indices has been discussed in Table 2.2.

For formulating the index, we have normalized the values of each component. As a result, the normalized values ranging between 0 and 1. “0” means the minimum while “1” means the maximum value. This procedure of normalization has been followed by the HDI methodology (UNDP, 2006).

All components have positive relationships through vulnerability which has been formulated accordingly.

$Y_{ij}=\frac{y_{ij}-\min(y_{ij})}{\max(y_{ij})-\min(y_{ij})}$, Min (y_{ij}) and Max (y_{ij}) are the values of “*j*”th indicator featuring low and high vulnerability of tribal communities. After normalization process, the averages of all the sub-indicators have been taken.

$$\text{ISE} = \sum (D_{\text{Pi}} + F_{\text{HHi}} + E_{\text{Ii}} + L_{\text{Hi}})/4 \tag{2.1}$$

Table 2.2 Formation of livelihood vulnerability index of the tribes

Tribal livelihood vulnerability index (TLVI)	Index of socio-economic (ISE)
	Index of livelihood (IL)
	Index of social security (ISS)
	Index of forest dependency (IFD)
	Index of health (IH)

$$\text{IL} = \sum (T_{\text{LAi}} + O_{\text{Hi}} + M_{\text{PCIi}})/3 \tag{2.2}$$

$$\text{ISS} = \sum (M_{\text{GNREGAIi}} + B_{\text{PDSi}} + S_{\text{SSi}})/3 \tag{2.3}$$

$$\text{IFD} = \sum (I_{\text{Fi}} + H_{\text{DFi}} + N_{\text{PCFi}})/3 \tag{2.4}$$

$$\text{IH} = \sum (H_{\text{HNTi}} + H_{\text{HNRTi}} + H_{\text{HNDWi}})/3, \tag{2.5}$$

where ISE means the index of socio-economic, D_{Pi} means the dependent people's percentage of the *i*th households, F_{HHi} means the households whether female headed or not, E_{Ii} means the average of the households' index of education level, L_{Hi} means the households' landholding, IL means the index of livelihood and T_{LAi} means the different type of livelihood activity by the *i*th households. O_{Hi} means the occupation of headed households, MPCIi means the per capita income of a month and ISS means the index of social security, M_{GNREGAIi} means the income generating from the scheme National Rural Guarantee Scheme, B_{PDSi} is the benefits received by households from the scheme public distribution system, S_{SSi} means the schemes of social security and IFD means the index of forest dependency, I_{Fi} means the forest income, H_{DFi} means the households who are dependent on forest, N_{PCFi} means the forest product collected (in number) and IH is the index of health, H_{HNTi} means the households who have toilet or not, H_{HNRTi} means households either received treatment from public hospital or not, H_{HNDWi} means households having drinking water or not. After normalization, we have taken the average of ISE, IL, ISS, IFD and IH to calculating livelihood vulnerability index of tribes.

For measuring climate variable index (CVI), three sub-indices have been used, P_{HHRRi} means the households' perception about rainfall shortage in the last five years, S_{Ii} means the households' perception for storm increased or not and N_{RWDi} means the households' perception about disaster warning increased or not (Table 2.3).

For formulating the index, we have taken the normalize values of all indicators. This normalization process was calculated by human development index (UNDP, 2006) methodology.

$$\text{CVI} = \sum (P_{\text{HHRRi}} + S_{\text{Ii}} + N_{\text{RWDi}})/3 \tag{2.6}$$

2.3.3 Analytical Model of Beta Regression

In order to estimate the factors of livelihood vulnerability of tribes, we have used the model of beta regression. As our dependent variable is the index of livelihood

Table 2.3 Components of climate variable index

Climate variable index (CVI)	Households' perception about rainfall shortage in the last five years	If yes = 1, others = 0
	Households' perception about storm increasing in the last five years	If yes = 1, others = 0
	Households' perception about not receiving disaster warning increased	If yes = 1, others = 0

vulnerability which also ranges between 0 and 1, we have used the model of beta regression.

We have assumed that the response variables are beta distributed; a model of beta regression (Cepeda-Cuervo, 2015).

$$g(\mu i) = \beta_0 + \beta_{1\mathrm{xi}1} + \beta_{2\mathrm{xi}2} + \beta_{3\mathrm{xi}3} + \beta_{4\mathrm{xi}4} + \beta_{5\mathrm{xi}5} + \beta_{6\mathrm{xi}6} + \beta_{7\mathrm{xi}7} = \eta_{\mathrm{i}},\ i = 1, \ldots, n \quad (2.7)$$

In the above equation, η_i means the linear predictor of ith households and $g(\mu)$ represents as the link function. The logit link has been used in this paper $[g(\mu) = (\log \mu/1 - \mu)]$ for beta regression, where Y_i is livelihood vulnerability index (LVI) of the ith household.

$X_{1\mathrm{i}}$ = Family size (in numbers).
$X_{2\mathrm{i}}$ = Agriculture income (in percentage).
$X_{3\mathrm{i}}$ = Distance from market (in kilometre).
$X_{4\mathrm{i}}$ = Distance from forest (in kilometre).
$X_{5\mathrm{i}}$ = Climate variable index.
$X_{6\mathrm{i}}$ = Time variable.
$X_{7\mathrm{i}}$ = Poverty scenario.

The model is heteroscedastic adjusted model. We have checked multicollinearity and autocorrelation and found that there is no such multicolliearity and autocorrelation (Tables 2.10 and 2.11 shown in the appendix).

2.3.4 *Results and Discussions*

The details of primary indicators and secondary indicators have been discussed in Table 2.4 based on literature of reviews and local tribal conditions.

The results of the index of livelihood vulnerability for the two time periods say 2015 and 2020 are shown in Table 2.5, and it is 0.440 in the year 2015, whereas it is 0.487 in the year 2020. From this table, we can see that the value of tribal livelihood vulnerability index has increased from the year 2015 to 2020. That means tribes are becoming more vulnerable over time.

Table 2.4 Details of all indicators of the index of livelihood vulnerability of tribes

Primary indicators	Secondary indicators	Description of secondary indicators
Index of socio-economy	Tribal peoples who are dependent	In % form
	Female headed tribal households	If yes = 1, others = 0
	Index of education	Average education
	Holdings of land	Acre
Index of Livelihood	Number of livelihood activity	Absolute terms
	Occupation of headed tribal households	Collection of NTFP = 1, mainly agriculture = 2, mainly labour (permanent and temporary) = 3, permanent service = 4, others = 5
	Monthly PCA	In Indian currency
Index of social security	MGNREGA income	In % to total income
	Benefits from PDS	If yes = 1, other = 0
	Social security schemes	Absolute terms
Index of forest dependency	Income from forest	In % to total income
	Households either dependent on forest	If yes = 1, others = 0
	Number of forest products collected	Absolute terms
Index of health	Households either having toilet facilities	If yes = 1, others = 0
	Households do not take healthcare facilities	If yes = 1, others = 0
	Households without having drinking water facilities	If yes = 1, others = 0

Source Field Survey

The tribal livelihood vulnerability indices over time with *t*-statistic have been given in Table 2.6. It has been studied from Table 2.6 that mean of livelihood vulnerability for 2015 is lower than 2020, and the mean difference is also significant. That means tribes are significantly becoming vulnerable over time in Kamo village of Bankura district.

On the basis of livelihood vulnerability, now, the tribes have been categories into two groups such as less and more vulnerable. Livelihood vulnerability of less vulnerable households is less than 0.50 while high vulnerable tribal households is beyond 0.50.

Figure 2.2 depicts the categorization of tribal communities who are less and more vulnerable in the village Kamo, Bankura district. About 74% households are less

Table 2.5 Livelihood vulnerability index of tribes of the village Kamo, Bankura district

Primary indicators	Secondary indicators	Score value	
		2020	2015
Index of socio-economy	Percentage of dependent people	0.612	0.386
	Female headed tribal households	0.140	0.040
	Educational index	0.394	0.483
	Land holding index	0.423	0.345
		0.392	0.313
Index of livelihood	Type of various livelihood activity	0.553	0.420
	Occupation of head	0.540	0.600
	MPCI	0.303	0.274
		0.465	0.431
Index of social security	Percentage income of MGNREGA	0.395	0.396
	PDS benefit	0.740	0.740
	Number of social security schemes	0.385	0.530
		0.507	0.555
Index of forest dependency	Percentage of forest income	0.494	0.579
	Household forest dependency	0.560	0.467
	Number of forest product collected	0.520	0.153
		0.525	0.400
Index of health	HHs do not have toilet	0.700	0.700
	HHs do not receive treatment	0.800	0.540
	HHs did not have drinking water facility	0.140	0.260
		0.547	0.500
Livelihood vulnerability index of the tribes		0.487	0.440

Source Authors' calculation

Table 2.6 Index of livelihood vulnerability of the tribes across two time periods (2015 and 2020) in the village Kamo, Bankura district

Year	Mean vulnerability index	*t* value	*P* value
2020	0.487138	2.581	0.011
2015	0.440007		

Source Authors' calculation

vulnerable and 26% households are more vulnerable in the year 2015 while 44% households are less vulnerable and 56% households are more vulnerable in the year 2020. This means that there has been an increase in tribal vulnerability over time. This further implies that the tribes are becoming more vulnerable over time.

In Table 2.7, livelihood vulnerability is the dependent variable, and the independent variables are the size of family, age of the headed household, the proportion

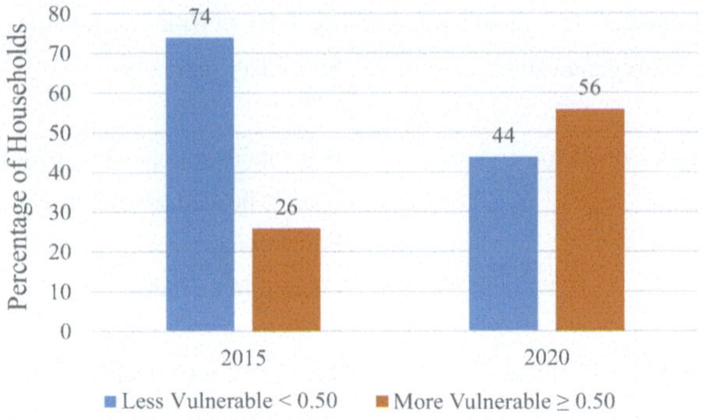

Fig. 2.2 % of less and more vulnerable of two time period

of income from agricultural, distance from home to market, distance from home to forest, climate variable index, time variable and poverty.

The estimated results of the model of beta regression of village Kamo in the Bankura district have been presented in Table 2.8. The model of beta regression is overall significant which has been showed by the value of $LRchi^2 = 140.34$ which means the above model of beta regression has a strong clarifying power. Among seven independent variables, six variables are significant statistically (Table 2.8). The result reflects that the coefficient of size of family is negatively associated with livelihood vulnerability which is statistically significant also. This signifies that the livelihood of tribes in Kamo is more vulnerable since there had been less household members. Further, if the members of family increases than the income of the family also rises which makes them less vulnerable. This result is also similar to a study of Philippine (Orbeta, 2005). From Table 2.8, it has been found that the coefficient of

Table 2.7 Basic statistics of dependent and independent variables

	Mean	Std. dev	Minimum	Maximum
Dependent variable				
Index of livelihood vulnerability	0.464	0.094	0.193	0.681
Independent variable				
Size of family (in numbers)	4.44	1.416	2	9
Age of head of household	49.98	14.02	24	90
Agriculture income (in percentage)	34.117	21.87	0	87.94
Distance from market (in kilometre)	0.674	0.299	0	1
Distance from forest (in kilometre)	7.02	0.173	6.5	8
Climate variable index	0.57	0.285	0	1
Time variable (2015 = 0, 2020 = 1)	0.5	0.503	0	1
Poverty* (BPL = 0, Other = 1)	0.08	0.273	0	1

Source Authors' calculation
* Poverty of the tribal households has been estimated by the procedure of Rangarajan Committee

agriculture income is negative and significant. Since agricultural income for tribes is less since they are more forest dependent that is why more forest income makes them less vulnerable. This result is also similar to the result of West Bengal in Paschim Medinipur district (Das & Basu, 2022). Table 2.8 shows that the coefficient of distance from home to forest is positively related with the vulnerability of tribes which is significant. That signifies that the family whose residences are near the forest is more vulnerable since they are more forest dependent. This result is quite similar with a study of Purulia district of West Bengal (Das, 2022). Table 2.8 shows that the coefficient of climate change is positive and significant with tribal vulnerability. That means that due to climate change, tribes are becoming more vulnerable (Deb & Mukherjee, 2022). The time coefficients of two time period say that 2015 and 2020 are positive and significant. This means that over time tribal community of Kamo village is becoming more vulnerable. Finally, the poverty is also negatively significant with tribal vulnerability. That means that poverty-ridden households are more vulnerable than APL households. This result is associated with Indonesia (Pritchett et al., 2000).

Sustainable local development strategies are taken by local people as adaptation strategies to solve the conundrum of achieving the sustainable development goals (SDGs). Tribal households have taken various adaptive sustainable local development strategies to minimize or to cop up climatic risks for achieving SDGs. Tribal people initially make adaptive capacity for risk minimization such as livestock rearing, forming SHGs, collection of NTFPs and seasonally migrating for more wage, etc. These adaptive strategies may boost them to enhance livelihood from the risks of climate.

Rearing of Livestock: The tribal families have domestic animals such as cows, hens, goats and pigs which help them to increase net revenue (Basu, 2017). The tribal

Table 2.8 Model of beta regression of the studied village

Model of beta regression

	Coeff	Std. error	*z*	$P > \|z\|$	Number of obs = 100LR Chi2(7) = 140.34 Prob > chi2 = 0.0000
Size of family (in numbers)	−1.199	0.608	−1.97	0.048	
Agriculture income (in percentage)	−0.155	0.033	−4.7	0.000	
Distance from market (in kilometre)	−1.455	2.359	−0.62	0.537	
Distance from forest (in kilometre)	6.221	0.497	12.51	0.000	
Climate variable index	6.853	2.473	2.77	0.006	
Time variable	4.100	1.633	2.51	0.012	
Poverty	−8.823	2.660	−3.32	0.001	
Constant	−5.448	0.062	−88.21	0.000	

Source Authors' calculation

Table 2.9 Different adaptation strategies in the Kamo, Bankura

Strategies of adaptation	Kamo
Rearing of livestock	20(40)
Formation of SHGs	22(44)
Accessibility of NTFPs	49(98)
Migration	22(44)

Note Figures in the parentheses is the % of sample households
Source Field survey in 2020

households rear some livestock and livestock products sold in the local market are hens, eggs, milk of cows, cow dung, goats' meat and pigs' meat. We have found that 40% households in the village of Kamo have adapted livestock rearing as alternative livelihood (Table 2.9).

Self-Help Groups (SHGs): SHGs have a major role to the tribal people for managing their livelihood from climatic changes. SHGs offer loan to the tribal women for building up their assets and alleviating poverty (Basu, 2017). From Table 2.9, it has been found that 44% households in the village of Kamo have formed SHGs.

Non-Timber Forest Products (NTFPs): Major NTFPs collected items are Sal leaves (used in making of sal dish), kendu leaves (for making biri), stones, various roots of the plants (used for medicine and marketing) and fuelwood (for consumption and marketing) (Basu, 2017). Tribal households also sell the forest products for their income generation. This study reveals that 98% households of Kamo are accessing NTFPs (Table 2.9).

Migration: In Kamo village, 44% tribal households are migrating for better wage. They generally migrate seasonally in the nearby developed districts such as Burdwan, Hooghly and Nadia of West Bengal, India.

2.4 Concluding Remarks

Firstly, from the two time period, the tribal communities in the year 2020 are more vulnerable (0.487) compared to that in the year 2015 (0.440). So, it has been observed that in the year 2020, tribal household are more vulnerable compared to the year 2015. The reasons are that there is no alternative source of income generating activities other than forest. Second, the livelihood vulnerability of tribes is influenced by the climate change, size of family, poverty scenario, agricultural income, distance from home to forest and time.

Sustainable local development strategies are taken by local tribal people as adaptation strategies to solve the conundrum of achieving the sustainable development goals (SDGs) through rearing of livestock, forming SHGs, collection of NTFPs and

Table 2.10 Correlation coefficient between dependent and independent variables

Variables	Tribal livelihood vulnerability index(TLVI)
Family size (in numbers)	−0.176*
Age of head of household	0.135*
Agriculture income (in percentage)	−0.384***
Distance from market (in kilometre)	0.021*
Distance from forest (in kilometre)	−0.084*
Climate variable index	0.254*
Time variable	0.252*
Poverty	−0.294**

Source Authors' calculation
Note * indicates significant at 1% level,** shows significant at 5% level and *** shows significant at 10% level

migrating for more wages. These strategies can help the tribes to enhance livelihood of achieving the sustainable development goals.

This paper recommends some policy interventions which will help to minimize the vulnerability of tribal community. Employment guarantee schemes can reduce vulnerability (Morduch, 1999). Social safety programmes help to minimize vulnerability (Devarajan & Jack, 2007). In Kamo village, agriculture is not so beneficial although major households still depend on it. So some steps have been given on agricultural development through irrigation facilities. Emphasis should also be given for non-farm employment, developing handicraft and small-scale industries and also formation of SHGs for livelihoods diversification of tribes. Further, anti-poverty programmes should be implemented efficiently for reducing poverty level of the tribal households.

Appendix

See Tables 2.10 and 2.11.

Table 2.11 Variation in factors of the variables

Variable	VIF	1/VIF
Climate variable index	1.21	0.825
Time variable	1.2	0.833
Agriculture income (in percentage)	1.14	0.876
Distance from forest (in kilometre)	1.13	0.881
Age of head of household	1.1	0.907
Family size (in numbers)	1.08	0.926
Distance from market (in kilometre)	1.06	0.946
Poverty	1.03	0.971
MeanVIF	1.12	

Source Authors' calculation

References

Adhikari, B. (2005). Poverty, property rights, and collective action: understanding the distributive aspects of common property resource management. *Environment and Development Economics, 10*(1), 7–31. https://www.jstor.org/stable/44378930

Ahsan, M. N., & Warner, J. (2014). The socioeconomic vulnerability index: A pragmatic approach for assessing climate change led risks-a case study in South-Western Coastal Bangladesh. *International Journal of Disaster Risk Reduction, 8*, 32–49.

Angelsen, A., Jagger, P., Babigumira, R., Belcher, B., Hogarth, N. J., Bauch, S., & Wunder, S. (2014). Environmental income and rural livelihoods: A global comparative analysis. *World Development, 64*(S1), S12–S28. https://doi.org/10.1016/j.worlddev.2014.03.006

Arnold, J. E. M., & Bird, P. (1999). Forest and the poverty-environment Nexus. In *Paper Presented at the UNDP-EC Expert Workshop on Poverty and the Environment, Brussels, Belgium.* http://citeseerx.ist.psu.edu/viewdoc/download?doi=10.1.1.526.1711&rep=rep1&type=pdf

Basu, J. P. (2017). *Climate change adaptation and forest dependent communities ananalytical perspective of different agro-climatic regions of West Bengal, India.* Springer.https://doi.org/10.1007/978-3-319-52325-5

Bhattacharya, P., & Prasad, R. (2009). Initial observation on impact of changing climate on NTFP resources and livelihood opportunities in Sheopur District of Madhya Pradesh (Central India). In *A Paper Presented at Xiii World Forestry Congress, Buenos Aires.* Argentina.

Cavendish, W. (1999). Poverty, inequality and environmental resources: Quantitative analysis of rural households. In *Oxford, Working Paper Series 99-9, Centre for the Study of African Economies.* https://econpapers.repec.org/paper/csawpaper/1999-09.htm

Cavendish, W. (2000). Empirical regularities in the poverty-environment relationship of rural households: Evidence from Zimbabwe. *World Development, 28*(11), 1979–2003. https://www.sciencedirect.com/science/article/abs/pii/S0305750X00000668

Census of India. (2011). Archived from https://censusindia.gov.in/2011-common/censusdata2011.html

Cepeda-Cuervo. (2015). Beta regression models: Joint mean and variance modeling. *Journal of Statistical Theory and Practice, 9.* https://doi.org/10.1080/15598608.2014.890983

Das, S. K. (2022). Forest dependency, social protection and tribal livelihood an empirical study in the backward district of Purulia, West Bengal. *SN Social Science, 2*(130). https://doi.org/10.1007/s43545-022-00431-7

Das, S. K., & Basu, J. P. (2022). Tribal livelihood vulnerability due to climate change: A study across tribes of Paschim Medinipur district of West Bengal. *SN Business Economics, 2*(103). https://doi.org/10.1007/s43546-022-00277-4

Deb, P., & Mukherjee, R. (2022). Household vulnerability of tribal people to climate change in the part of Dooars region, West Bengal, India. In *Regional development planning and practice*. Print ISBN: 978-981-16-5680-4, Electronic ISBN: 978-981-16-5681-1. https://doi.org/10.1007/978-981-16-5681-1_15

Devarajan, S., & Jack, W. (2007). Protecting the vulnerable: The trade off between risk reduction and public insurance. *World Bank Economic Review, 21*(1), 73–91.

FAO. (2006). Global forest resources assessment-progress towards sustainable forest management. In *FAO Forestry Paper 147*. Rome: Food and Agriculture Organisation of the United Nations. http://www.fao.org/3/a0400e/a0400e00.htm

Fisher, M. (2004). Household welfare and forest dependence in Southern Malawi. *Environment and Development Economics, 9*, 135–154.

Ghosh, M., & Ghosal, S. (2020). Climate change vulnerability of rural households in flood-prone areas of Himalayan foothills, West Bengal, India. *Environment Development Sustainable*. https://doi.org/10.1007/s10668-020-00687-0

Hahn, M. B., Riederer, A. M., & Foster, S. O. (2009). The livelihood vulnerability index: A pragmatic approach to assessing risks from climate variability and change: A case study in Mozambique. *Global Environmental Change, 19*(1), 74–88.

INCCA. (2010). Climate change and India: A 4X4 Assessment. Government of India. http://www.indiaenvironmentportal.org.in/files/fin-rpt-incca.pdf

IPCC. (2007). Climate change 2007: the physical science basis. In S. Solomon, D. Qin, M. Manning, Z. Chen, M. Marquis, K. B. Averyt, M. Tignor, & H. L. Miller (Eds.), *Contribution of Working Group I to the Fourth Assessment Report of the IPCC*. Cambridge: Cambridge University Press.

Malhotra, K. C., & Bhattacharya, P. (2010). Forest and Livelihood. CESS Publisher, Hyderabad, India, 246. https://scialert.net/abstract/?doi=rjf.2016.1.7

Morduch, J. (1999). Between the state and the market: Can informal insurance patch the safety net? *World Bank Research Observer, 14*(2), 187–207.

Mukul, S. A., Rashid, A. Z. M. M., Uddin, M. B., & Khan, N. A. (2016). Role of non-timber forest products in sustaining forest-based livelihoods and rural households' resilience capacity in and around protected area: A Bangladesh study. *Journal of Environment Planning and Management, 59*(4), 628–642. https://doi.org/10.1080/09640568.2015.1035774

Orbeta, A. Jr. (2005). *Poverty, vulnerability and family size: Evidence from the Philippines*. Philippine Institute for Development Studies. https://www.researchgate.net/publication/24110971_Poverty_Vulnerability_and_Family_Size_Evidence_from_the_Philippines

Pandey, A. K., Kori, D. C. (2011). Non-destructive harvesting practices of Terminaliaarjuna (Arjuna), Phyllanthusemblica (Aonla) and Andrographispaniculata (Kalmegh). *Indian Forester, 137*, 1269–1279. https://www.researchgate.net/publication/285855238

Pritchett, L., Suryahadi, A., & Sumarto, S. (2000). Quantifying vulnerability to poverty: A Proposed measure, applied to Indonesia. In *Policy Research Working Paper; No. 2437*. Washington, DC: World Bank. https://openknowledge.worldbank.org/handle/10986/21355

Reddy, S. R. C., & Chakravarty, S. P. (1999). Forest dependence and income distribution in a subsistence economy: Evidence from India. *World Development, 27*(7), 1141–1149. https://doi.org/10.1016/S0305-750X(99)00057-1

Saxena, N. C. (2003). Livelihood diversification and non-timber forest products in Orissa: Wider lessons on the scope for policy change? In *ODI Working Paper No. 223*. London, UK: Overseas Development Institute. https://cdn.odi.org/media/documents/2460.pdf

Shackleton, C. M., Pandey, A. K., & Ticktin, T. (2015). *Ecological sustainability for non-timber forest products: Dynamics and case studies of harvesting*. Routledge.

Thomas, D. S. G., & Twyman, C. (2005). Equity and justice in climate change adaptation amongst natural-resource-dependent societies. *Global Environmental Change, 15,* 115–124. https://doi.org/10.1016/j.gloenvcha.2004.10.001
UNDP. (2006). Human development report, United Nations Development Program. http://hdr.undp.org/sites/default/files/reports/267/hdr06-complete.pdf

Chapter 3
Role of Fuelwood in Sustainable Livelihood Security of Tribal Community of Kinnaur District, Himachal Pradesh, India

Rajan Maurya, Ganesh Yadav, B. W. Pandey, and V. S. Negi

Abstract Indian Himalayan region has a vast area of forest resources. Different types of forests are available in the lap of the Himalayas which differs with other mountain systems in the world. Kinnaur is one of the most forest-covered districts of Himachal Pradesh in the western Himalayan order. Rich forest area is the source of abundant forest products which leads to sustain the livelihood of tribal community and associated rural economy. This study shows the availability of forest products in the valley of Nichar, Kalpa and Poo and impact of availability of forest resources on the livelihood security of Kinnauras, the main tribal community of Kinnaur. It includes the local traditional knowledge and practices of the tribal people to sustain their livelihood practices. This study has been based on primary as well as secondary data sources. Different statistical techniques have been used in STATA 16.0 software to analyze the data. This study develops a result that rapid developmental activities in the valley have changed the local environment which affects the sustainable livelihood of the local people and role of local governments in the well-being of the study area.

Keywords Indian Himalayan region · Rural economy · Sustainable livelihood · Local traditional knowledge · Tribal community

R. Maurya (✉)
Department of Geography, Indraprastha College for Women, University of Delhi, Delhi 110054, India
e-mail: rajanmaurya@ip.du.ac.in

G. Yadav
Department of Geography, Kalindi College, University of Delhi, Delhi 110008, India

B. W. Pandey
Department of Geography, Delhi School of Economics, University of Delhi, Delhi, India

V. S. Negi
Shaheed Bhagat Singh College (Ev.), University of Delhi, Delhi, India

© The Author(s), under exclusive license to Springer Nature Switzerland AG 2024

B. V. Raj Singh and A. K. Batar (eds.), *Sustainable Local Development for Environmental and Social Sustainability*, Human-Environment Interactions 11,
https://doi.org/10.1007/978-3-031-67303-0_3

3.1 Introduction

Human beings have continuously modified their habitat from ancient times according to their needs and uses, usually with forests land giving way to agriculture, livestock grazing and urbanization in different parts of the world. The changing pattern of needs of human beings changed subsistence livelihoods which brings a certain degree of change in land use inevitable and acceptable in the interests of social and economic progress (Chakraborty et al., 2009). The forest ecosystem is very fragile system that serves both nature and humans for their sustainable livelihood. But, in the present time of changing world, forest-related services to human beings are affected by the forest ecosystem's degradation. Different developmental activities have changed the scenario of the local environment which affects the forest ecosystem and associated services for the livelihood of tribal community in mountainous region of the world. The same thing happened on the roof of the world's Himalayas which is also known as the water tower of world.

A healthy and well-protected forest can complete the present and future needs and well-being of human beings in the wake of different cultural, aesthetical products and services. Forest is the primary source of the livelihood of the tribal communities worldwide; in other words, we can say that forest is the backbone of the indigenous community's livelihood and prosperity. The analysis of wealth in livelihood to understand rural local inequalities has received considerable focus during the last few decades in developing countries and India (Sharma, 2005). Sustainable livelihood can cope and recover from stress and shocks of, maintain or enhance its original capabilities, assets and entitlements, both current and future needs, while not undermining the actual natural resource availability (Chamber & Conway, 1992). However, the scale of the reaction looks insufficient to avert catastrophic forest damage caused by present fuelwood gathering techniques (Heltberg et al., 2000).

The relevance of farmer participation in the evaluation of fodder features in the creation of better dual-purpose varieties is highlighted by these findings. The impact of these types on impoverished farm households, however, will be dependent on a corresponding increase in seed system efficacy (Rao & Hall, 2003).

Indigenous knowledge is an essential part of the Himalayan region for sustainability and well-being. It is like intergenerational wisdom of local tribal people to toward their security of livelihood, culture, practices in a most sustainable manner under the harsh conditions of climate, isolation, inaccessibility, remote locality and modern connectivity. This traditional knowledge is transferred from one generation to the next one by spoken words. It is scientific in traditional experimentation and dynamic in dispersal and receiving regular impetus from the outside world.

3.2 Study Area

Kinnaur is situated in the western part of the Indian Himalayan and is one of the twelfth administrative districts of Himachal Pradesh; longitudinal and latitudinal extents are between 77° 45′ and 79° 00′ 35″ East, 31° 05′ 50″ and 32° 05′ 15″ North. Kinnaur is divided into three development blocks, i.e., Nichar, Kalpa and Pooh. The administrative headquarter of Kinnaur district is at Reckong Peo. According to forest statistic report, geographical area of Kinnaur is 6401 km^2 which has 79.81 km^2 very dense forest, 329.28 and 236.90 km^2 moderately dense forest and open forest, respectively. It has total of 645.99 km^2 of forest cover which is about 1.08% of the states' geographical area. It is one of the scheduled tribe districts of Himachal Pradesh and the tribes are dependent on the forest ecosystem and its services for their livelihood options (Fig. 3.1).

3.3 Objective of Study

(a) To identify the role of forest products in the sustainable livelihood of tribal community.
(b) To analyze the developmental activities and their impact on local environment.

3.4 Material and Methods

The study is based on primary and secondary data sources. For primary survey data collection, ten villages have been taken from different forest ranges of Kinnaur on the basis of proximity of the nearest forest. It has been taken to identify the different forest products and dependency of the tribal community on it. Extensive field survey has been conducted with the help of pre-structured and pre-tested questionnaire, non-participant observation of study area, and scheduled interview for obtaining primary data on the uses of forest products and the impact of different developmental activities on the local environment. For their selection, stratified random sampling (SRS) has been incorporated to take households from the different villages.

For the fulfillment of objectives, household survey has been done in the villages that fall in the proximity of forest to understand the role of forest products on sustainable livelihood security. The strata for the selection process of villages are done on the basis of height i.e. below 2500 m to above 3500 m. Household survey has been done from the 150 households from the selected villages. With the help of pre-tested questionnaire, consumption of fuelwood by households per day in different seasons, uses and collection of different parts of the medicinal plants have been identified for the study. Apart from the fuelwood, consumption of fodder by livestock which includes crop residue, dry and green grasses and some dry leaves of Chilgoza pines has been incorporated. The questionnaire survey data has been compiled on MS

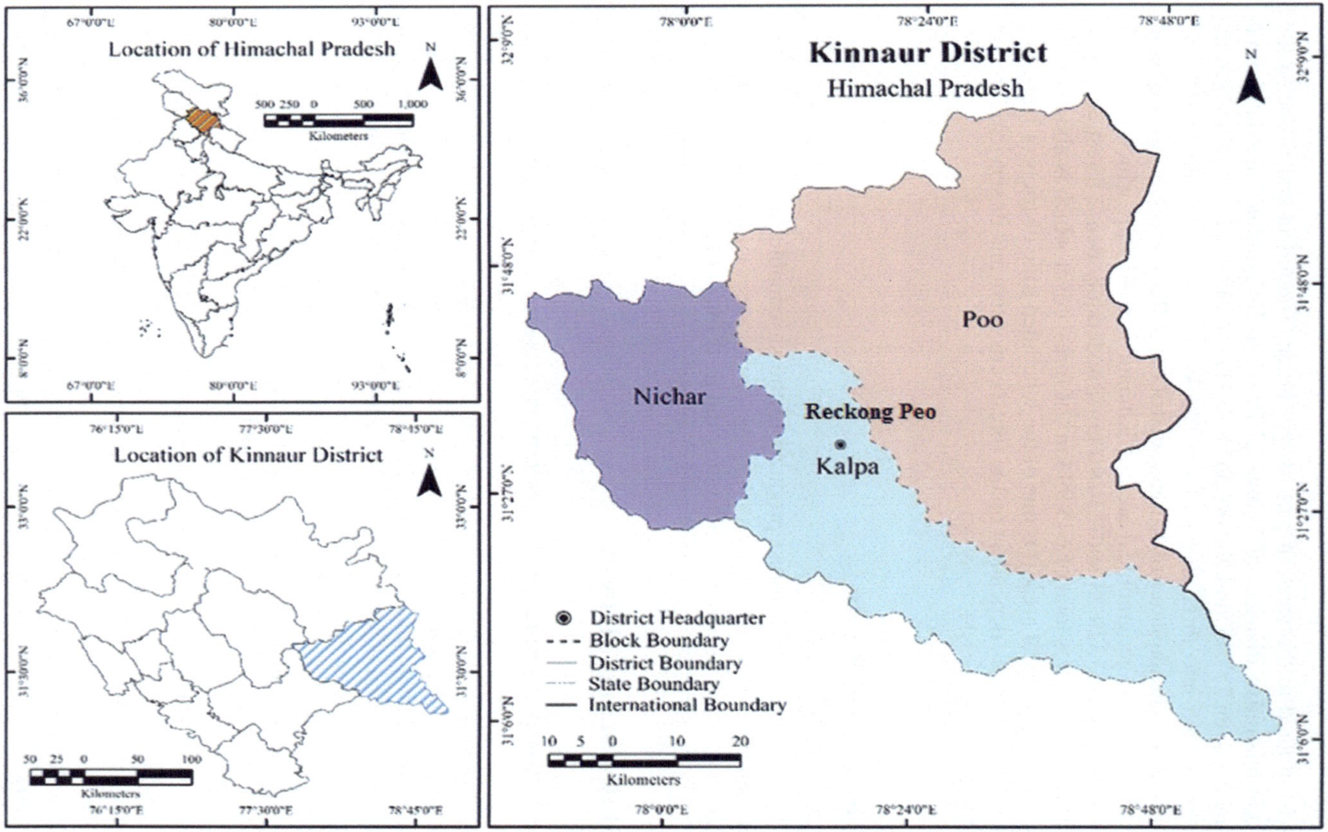

Fig. 3.1 Map of the study area: Kinnaur. *Source* Prepared by the Author on Arc GIS, 2022

Excel and processed in STATA 16.0 for the representation of results through various graphs, charts and tables.

3.5 Result and Discussions

According to World Bank (2006), almost 1.6 billion people depend on forest products for livelihood, and 350 million live within forest proximity, out of which 60 million people are the local tribal population which is dependent on forest product resources directly or indirectly for their sustainable livelihood. According to Government of India report, 1.73 Lakh villages live in the forest area proximity for their livelihood (MoEFCC, 2015). Most local tribal people living in the forest area derived up to 60% of the economic by selling different forest products without counting domestic consumption. Over 70% of the rural tribal population of India directly derived their source of energy from the forest, and 100 million individuals depend on forest products for livelihood (Raj et al., 2020). The tribal community depends on various activities of the forest for income generations. There is a great scope for the tribal community to develop using all the forest resources products sustainably. People who live near the forest area or live inside the forest area have more economic gain. The better infrastructure or connectivity could accelerate the price and revenues of the forest products.

3.6 Major Tribes of Kinnaur

In Himachal Pradesh, the tribal population accounts for 4.61% of the total population. Within the state of Himachal Pradesh, Kinnaur is one of the highest concentrations of tribal population, having 57.95% of the total population spread over 77 villages. Tribes residing in Kinnaur are Bodh, Gaddi, Gujjar, Jad, Lamba, Khampa, Kanaura, Kinnara, Lahaula, Pangwala and Swangla.

3.7 Nature of Tribal Livelihood

Agriculture constitutes the primary source of livelihood among tribes in India. It plays a vital role in the national economy, rural development, employment and occupation, agro-industries, food and nutrition security, growth and survival, social, economic and cultural conditions, and poverty alleviation (Surayya et al., 2008). About 70% of the population mainly depends on rain-fed agriculture characterized by low productivity, unpredictive weather and calamities, degraded soil with low fertility, unprotective irrigation and degraded natural resources (Chakraborty et al., 2009). The traditional livelihood system of tribal people of Kinnaur was based on the complex set of social and economic with households and their village community. The living situation of tribal communities has been characterized by the mountains they have been adapted by traditional survival strategies (Jodha, 1990a, b, 1997). The tribal community is believed to adapt to the severely changing environment, including livelihood diversification, social and capital (Goulden et al., 2013).

3.8 Forest Products as Livelihood Options for Tribal People of Kinnaur

Livelihood refers to capabilities, assets (both material and social resources) and activities which become the means for human living (Chambers & Conway, 1992). The tribal communities in Kinnaur have been closely dependent on forest as their livelihood means since ages. Changes in the forest cover have negatively affected the forest-based livelihood options. Main forest products are used by the local community for their livelihood options.

Fuelwood—It is the integral part of the local ecology and their traditional domestic fuelwood system.

Fodder—Collection of dry leaves, dry grasses and crop residue from the forest and agriculture field to meet their need.

3.9 Uses of Fuelwood

The fuelwood is being used by the poor in many parts of the developing countries, likewise people of Kinnaur also depend on fuelwood for domestic cooking energy and heating of the homes during the winter season to protect themselves from cold (Muneer, 2003; Pine et al., 2011). The use of fuelwood is driven by the high availability of woods in forest, available at no cost, good quality of woods present, lack of alternatives and infrastructure. As per primary data survey, the respondents have responded that each and every household of Kinnaur consumes fuelwood as a source of energy.

Table 3.1 states that every household needs fuelwood for cooking as well as the heating of the houses during the winter season, and this fuelwood is to be collected from the forest. In winter season, there is always a scarcity of fuelwood and has to be collected during the summer season and stored for winter season. It has been found that consumption of fuelwood per day is 1.80 kg in summer for households of Poo followed by Nichar, Malling and Kilba. Poo also has the highest percentage of per capita consumption of fuelwood per day/kg in winter (6.91%) followed by Nichar, Malling and Kilba. The lowest consumption is found in Kalpa followed by Bhaba Nagar and Katgaon (Table 3.1). This is because Poo and Malling are at the

Table 3.1 Forest range-wise consumption of fuelwood

Forest range name	Per capita fuelwood consumption per day in kg during summer	Per capita fuelwood consumption per day in kg during winter	Per capita fuelwood consumption per day annual in kg	Total fuelwood consumption per household per day in summer in kg	Total fuelwood consumption per household per day in winter in kg
Bhaba Nagar	1.52	6.03	3.78	7.83	31.12
Kalpa	1.46	5.73	3.60	6.95	27.30
Katgaon	1.60	6.16	3.88	7.77	29.88
Kilba	1.65	6.47	4.06	6.98	27.37
Moorang	1.58	6.10	3.84	6.77	25.75
Malling	1.65	6.53	4.09	8.03	31.71
Nichar	1.71	6.70	4.20	8.10	31.81
Poo	1.80	6.91	4.35	6.98	26.81

Source Primary Survey, 2020

upper part of Kinnaur where the temperature remains relatively low as compared to other areas. Nichar and Kilba also have higher consumption because the distribution of forest is very high so there is abundant availability of fuelwood. But the per day consumption of fuelwood in kg is always greater in winter season for all the forest ranges as the temperature drops down below minus degree Celsius and not only for cooking purposes but they are also used for heating of their houses to keep them warm. In the summer season, fuelwood is only used for cooking purposes.

3.10 Uses of Fodder

One of the major occupations of the local people of Kinnaur is livestock rearing, and it is an integral part of economy; this livestock depends on fodder which is obtained from agricultural residue and agro-forestry greens, greens leaf of woods collected from forest (Purohit & Samant 1995; Singh, 1990).

Table 3.2 states that fodder is one of the most important elements for the rearing of livestock animals (cow, yak, goat and sheep) and farmer mostly depend from agro-forestry; as well as green and dry fallen tree leaves have been collected from the forest, whereas the green fodder is necessary for the production of dairy products. It has been found that Kilba has maximum use of fodder at 21.08% followed by Kalpa (18.65%), Nichar (14.59%) and Malling (12.5%). The minimum use of fodder has been found at Moorang (7.03%) and Poo (8.11%). While the percentage of not using the fodder has been almost same in both Katgaon (21.43%) and Poo (14.29%) forest ranges. Some of the households of Bhaba Nagar and Moorang have not used the fodder because they do not have any kind of livestock (Table 3.2). They are generally

Table 3.2 Forest range-wise percentage use of fodder

Forest ranges	Don't use fodder	Use fodder
Bhaba Nagar	0	8.11
Kalpa	7.14	18.65
Katgaon	21.43	9.73
Kilba	28.57	21.08
Malling	7.14	12.7
Moorang	0	7.03
Nichar	21.43	14.59
Poo	14.29	8.11
Total	100	100

Source Primary Survey, 2020

involved in government jobs, hotels and restaurants for their livelihood means. The Northwestern part of Kilba has the highest percentage of household where fodder is used because this range has one of the largest stretches of forest land where tree fodder is easily available and settlement is also higher so the livestock percentage is higher here. The region with more settlement area has higher agricultural practices which have been performed by the local tribal people. The residue of crops has also been used as fodder in the area where agricultural practices are performed by the locals (Plate 4.4). The southern part of Kilba has less settled and forest area is also dispersed so the use of fodder for livestock by the households is less. Moorang has no household using fodder because the area has diversifying livelihood opportunities like apple orchards cultivation, business, hotels, restaurants, etc.

3.11 Spatial–Temporal Analysis of Availability of Different Forest Products

Spatial–temporal variability of forests is becoming the role of environment and human society. Spatial variation within the region shows the communities develop and involves size, pattern, nature and quantity of different types of forest while the temporal becomes the changes duration that brings changes to them. During the different time periods, the structure and nature of the forests have changed, whereby some forests are converted to open forest and vice-versa (afforestation). When we look at time and spaces of forests, there is a great influence of human society. The locals have been depending on forests for livelihood extracting different forest products like cutting down or collecting woods for fuelwood, tree leaves for livestock as fodder, medicinal plants for curing diseases and pasture land for grazing. But this availability has changed from time to time according to season as fuelwood and fodder are in shortage during winter but are abundance in summer season.

Table 3.3 Time spent on the collection of fuelwood: (percentage of time spent on fuelwood h/day)

Forest ranges	No visit	Below 3 hrs	3–5 hrs	05–10 hrs	Above 10 hrs
Bhaba Nagar	7.32	51.06	2	0	0
Kalpa	12.2	12.77	26.67	23.17	0
Katgaon	4.88	6.38	22.67	0	0
Kilba	29.27	12.77	28	26.83	0
Malling	17.07	0	0	17.07	42.19
Moorang	7.32	0	0	7.32	26.56
Nichar	14.6	17.02	20.67	13.41	1.56
Poo	7.32	0	0	12.2	29.69
Total	100	100	100	100	100

Source Primary Survey, 2020

3.12 Time Spent on Collection of Fuelwood by Tribal People

The consumption rate of fuelwood is high in Kinnaur as majority of the local community depends on it. It needs to be collected everyday as per their need and consumption patterns. The local people need to spend hours in collecting the fuelwood selecting the dry woods. The hours may vary depending upon the types of forest like open forest may take longer time while in the dense forest, it takes less time as the woods are available in abundance. Also, there is season-wise variation for time spent on collecting it (Table 3.3).

3.13 Changes in Availability of Fuelwood

Under the development process, the availability of fuelwood has changed, and the forest area is converted into apple orchards, walnuts and almond plantations. Agricultural practices are common where the good quality of fuelwood is declining, which requires too much effort to collect it. There are many reasons for the changes in the availability of fuelwood, and even seasonal changes also affect the availability of fuelwood (Table 3.4).

3.14 Time Spent on Collection of Fodder

There are two options to obtain fodder for their livestock: one is from agro-forestry product and another is green leaf from the forest. Either collecting from the agriculture field or from the forest they need to pluck leaves, stems and arrange which takes

Table 3.4 Forest range-wise changes in availability of fuelwood

Forest ranges	Increase	Decrease	Can't say
Bhaba Nagar	10.47	8.04	4.05
Kalpa	12.79	16.96	28.38
Katgaon	17.44	8.04	8.11
Kilba	11.63	23.66	25.68
Malling	10.47	15.18	6.76
Moorang	12.79	5.36	4.05
Nichar	11.63	16.52	13.51
Poo	12.79	6.25	9.46
Total	100	100	100

Source Primary Survey, 2020

around 3 h depending on the number of livestock locals have and amount of fodder they are collecting. If the number of livestock is more, then they have to spend longer period of time.

Table 3.5 shows that the availability of fodder in a confined place would reduce the time spent to collect the fodder for livestock but this was untrue in many of the eight forest ranges of the study area. Study has found that the highest number of households that do not spend time on collection of fodder is 29.27% from Kilba, Malling, Nichar and Kalpa. The lowest percentage (7.32%) belonged to Bhaba Nagar followed by Moorang.

It was found that the highest percentage of households (29.55%) from Nichar spent less than 3 h to collect fodder followed by Kilba (21.21%), Bhaba Nagar (20.45%) (Table 3.5). It can be seen that the highest percentage of households (40.45%) from Kalpa spent 3 to 5 h to collect fodder for livestock followed by Katgaon (17.12%), Kilba (15.32%) and Malling (12.61%). The lowest percentage of households (5.41) is from Moorang and Poo (9.01%). The highest percentage of households (27%) of

Table 3.5 Forest range-wise time spent on collection of fodder

Forest ranges	No visit	Below 3 hrs	3–5 hrs	05–10 hrs
Bhaba Nagar	7.32	20.45	0	0
Kalpa	12.2	15.15	40.54	0
Katgaon	4.88	13.64	17.12	0
Kilba	29.27	21.21	15.32	25
Malling	17.07	0	12.61	27
Moorang	7.32	0	5.41	17
Nichar	14.63	29.55	0	12
Poo	7.32	0	9.01	19
Total	100	100	100	100

Source Primary Survey, 2020

Malling have spent less than 5 to 10 h to collect fodder followed by Kilba (25%), Poo (19%) and Moorang (17%) because they use good amount of agro-residue, the mountains are gentle slopes and forests are available at nearby places. Malling and Moorang have recorded more time beyond 3 h because this area comes under rain shadow zone of the district, some parts of the barren land and some of these lands have been converted into apple orchards by the local people for completing the demand of the apples in the national market. Mountains in this forest range have very steep slopes which are not favorable for the collection of fodder by the local people in rainy and winter seasons.

3.15 Changes in Availability of Fodder

Fodder availability has change in recent times as the living standards of locals have changed and the number of livestock has increased but the supply of fodder (dry fodder from agricultural residue, green leaf from forest tree) has reduced because of forest area encroachment and apple orchards cultivation.

Table 3.6 shows the availability of fodder for livestock depending on seasons, forest resources and agricultural crop residue. It has been observed that the highest percentage of the households of Malling (31.52%) said that there is an increased incidence of scarcity of fodder because of the increasing number of developmental activities and growing horticulture. About 22% and 19.57 households of Poo and Moorang have complained about the change in fodder availability. It has been found that the lowest percentage is in Kilba (1.09%) and Bhaba Nagar (3.26%). It has also been observed that the highest percentage of about 26.05% of household from Kilba said that there is a decrease in the changes in availability of fodder followed by Kalpa (24.65%), Nichar (18.14%) and Katgaon (12.09%).

In contrast, the lowest percentage is available at Poo, Moorang and Malling, i.e., 1.86%, 2.79% and 4.65%, respectively (Table 3.6). Malling, Poo and Moorang have

Table 3.6 Forest range-wise changes in availability of fodder

Forest ranges	Increase	Decrease	Can't say
Bhaba Nagar	3.26	9.77	7.79
Kalpa	7.61	24.65	12.99
Katgaon	6.52	12.09	9.09
Kilba	1.09	26.05	32.47
Malling	31.52	4.65	11.69
Moorang	19.57	2.79	2.6
Nichar	7.61	18.14	14.29
Poo	22.83	1.86	9.09
Total	100	100	100

Source Primary Survey, 2020

seen the shifting of forest area into the cultivation of apple orchards, and construction of big hotels, restaurants, dams and tree line shifting upwards shows the increased change in the availability of fodder. Kilba, Kalpa, Nichar and Katgaon have a great stretch of lush green forest. The forest is also increasing simultaneously with other activities, which let local people say there is a decrease in the availability of fodder in their area. Upper part of Kinnaur sees much increase while lower altitudes do not see such increase as this area comes under rain shadow due to the Dhauladhar range.

3.16 Impact of Different Forest Products on Livelihood

The uses and extraction of different forest products like fuelwood, fodder, medicinal plants and grazing have impacted the livelihood of locals. This impact could be positive or negative; positive impact could be extraction of forest products have protected from alien plants but it would affect the nature of the local ecosystem which leads to unproductivity of locals' livelihood and change in local micro-climate, over utilization of forest products can also limit access to resources.

3.17 Fuelwood

Consuming fuelwood has major impact to the livelihood of locals as many of them were women who manage cooking and heating of homes which created health issues-related problems. The collection from the forest and traveling for more than 3–4 h daily carrying on their head may lead to health problem like back pain, leg joint problems, etc.

It was found that the highest percentage of households (21.69%) of Kilba has responded that there is an impact of fuelwood on livelihood followed by Kalpa (18.38%), Nichar (14.34%) and Malling (11.76%), while the lowest percentage of households (7.35%) is from Moorang (Table 3.7). It has been observed that Kilba and Kalpa have the highest percentage of household impacted by fuelwood on their livelihood because carrying a fuelwood on their head for long distances more than 6 km every day for longer period would result in health problem, affected in spinal cord (back pain) joint pain, and continuous burning of fuelwood would result in lung problem. Spending almost all day in forest collecting fuelwood does not allow them for other household activities, etc.

Table 3.7 Forest range-wise fuelwood impact on livelihood

Forest ranges	No impact	Impact	Can't say
Bhaba Nagar	9.78	7.72	0
Kalpa	15.22	18.38	30
Katgaon	7.61	10.29	20
Kilba	22.83	21.69	10
Malling	11.96	11.76	25
Moorang	4.35	7.35	10
Nichar	18.48	14.34	5
Poo	9.78	8.46	0
Total	100	100	100

Source Primary Survey, 2020

3.18 Fodder

Green fodder and dry agricultural/crop residue are the major fodder for livestock. The use of this has impacted the livelihood of locals as milk production increased if the nutritious fodder is fed and it might give best breed. It could also uplift the social structure for the locals (Pandey et al., 2021).

From the Table 3.7, it is evident that 20% of households of Kilba have found that there is no impact of fodder on livelihood followed by Malling and Katgaon, i.e., 15% and 14.17%, respectively (Table 3.8). At the same time, the lowest percentage is observed at Moorang (3.33%) then Bhaba Nagar (8.33%). It has been found that the highest percentage of households (21.67%) from Kilba have responded there is an impact of fodder on livelihood. About 19.58, 15.42 and 11.67% of households from Kalpa, Nichar and Malling also observed an impact of fodder on livelihood while the lowest percentage is recorded at Poo (6.25%) and Bhaba Nagar (7.5%). Kilba, Kalpa and Nichar have wide variety of fodder like the herbs, medicinal value with economic value. If nutritious fodder is given, then it will give profit to milk production. Poo has the lowest impact by fodder on livelihood because of cold deserted barren region, high altitude, steep slope and less population of cows.

3.19 Degree of Impact of Forest Cover Change on Livelihood

The degree of impact of forest cover change on livelihood is the intensity of changes that impacted by the changes in forest cover due to the different activities happening on forest like extraction of different forest products for livelihood (Maurya et al., 2021).

Table 3.8 Forest range-wise impact of fodder on livelihood

Forest ranges	No impact	Impact	Can't say
Bhaba Nagar	8.33	7.5	8.33
Kalpa	12.5	19.58	33.33
Katgaon	14.17	8.75	4.17
Kilba	20	21.67	25
Malling	15	11.67	8.33
Moorang	3.33	9.17	0
Nichar	15	15.42	8.33
Poo	11.67	6.25	12.5
Total	100	100	100

Source Primary Survey, 2020

Figure 3.2 shows the degree of impact of forest cover change on livelihood. We have calculated the degree of impact of forest cover change on livelihood and the impacts were recorded formulating into four different categories—not impacted, slightly impacted, impacted and highly impacted. It has been evident that the highest percentage of households (20.69%) of Malling said they were not impacted by the forest cover change followed by Kilba, Nichar and Malling. It also has been found that the lowest percentage of households (10.34%) of Poo was not impacted by the degree of forest cover change on livelihood succeeded by Moorang, Katgaon and Bhaba Nagar (Fig. 4.4).

About 20% of households of Kalpa were slightly impacted followed by Katgaon, Moorang and Poo which were also slightly impacted. Households of Nichar, Malling and Bhaba Nagar have also been impacted by the degree of forest cover change

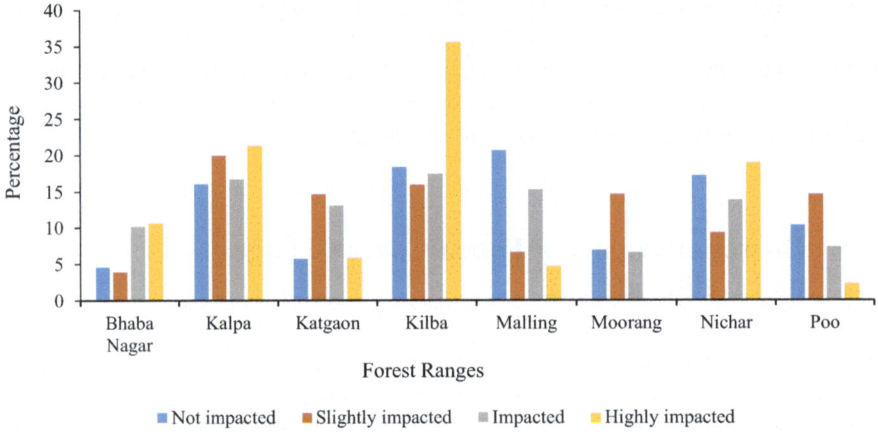

Fig. 3.2 Forest range-wise degree of impact of forest cover change on livelihood. *Source* Primary Survey, 2020

on livelihood. About 17.39% and 35.71% of households of Kilba have the highest percentage of households being impacted and highly impacted by the degree of change of forest cover on livelihoods followed by Kalpa. It has been evidenced from the table that 10.71% of households of Bhaba Nagar are only impacted with the forest cover change over a period of time.

3.20 Conclusion

It is obvious that the non-timber forest products contribute a lot to providing livelihood and income generation of the locals' tribal living near the forest. A large number of local tribal community derive a number of forest products like fuelwood as a source of energy, fodder for livestock, medicinal plants for treatment, other fruits, woods for house construction. They also have a diversifying livelihood option like horticulture. Apple cultivation is one of the main occupations of the local tribal people as the soil and geological structure support them. Tourism and adventurous sports are also taking up as an occupation by the young locals for their livelihood.

The increasing livelihood of local tribal community dependent on forests requires some focused interference on NTFPs. Facilities per to value, storage, of programs and schemes in both private and public sectors should be created and promoted. The local tribal community should be informed about products, market, policy to enable them to utilize and have proper access and returns from NTFPs.

References

Chakraborty, P., Tiwari, H. R., & Jha, M. K. (2009). Sustainable rural livelihood through participatory natural resource management: A case study. *Journal of Rural Development, 28*(1), 85–100.

Chambers, R., & Conway, G. (1992). *Sustainable rural livelihoods: Practical concepts for the 21st century*. Institute of Development Studies (UK).

Chambers R, Conway G, & Brighton Institute of Development Studies. (1992). *Sustainable rural livelihoods: practical concepts for the 21st century* (Vol. 296, pp. 1–29). Brighton: Institute of development studies.

Goulden, M. C., Adger, W. N., Allison, E. H., & Conway, D. (2013). Limits to resilience from livelihood diversification and social capital in lake social-ecological systems. *Annals of the Association of American Geographers, 103*(4), 906–924. https://doi.org/10.1080/00045608.2013.765771

Heltberg, R., Arndt, T. C., & Sekhar, N. U. (2000). Fuelwood consumption and forest degradation: a household model for domestic energy substitution in rural India. *Land Economics*, 213–232.

Jodha, N. S. (1990a). Mountain agriculture: The search for sustainability. *Journal of Farming Systems Research Extension, 1*(1), 55–75.

Jodha, N. S. (1990b). Rural common property resources: contributions and crisis. *Economic and Political Weekly*, A65-A78.

Jodha, N. S. (1997). 20 agricultural growth and sustainability: perspectives and experiences from the himalayas. *Sustainability, Growth, and Poverty Alleviation: A Policy and Agroecological Perspective*, 294.

Maurya, R., Negi, V. S., & Pandey, B. W. (2021). Spatio-temporal analysis of land use/land cover change through overlay technique in Kinnaur district of Himachal pradesh, Western Himalaya. *Sustainability, Agri, Food and Environmental Research, 9*(1), 50–65.

Ministry of Environment, Forest and Climate Change (MoEFCC). (2015). India: First Biennial Update Report to the United Nations Framework Convention on Climate Change.

Muneer, S. E. T. (2003). Adoption of biomass improved cookstoves in a patriarchal society: An example from Sudan. *Science of the Total Environment, 307*(1–3), 259–266.https://doi.org/10.1016/S0048-9697(02)00541-7

Pandey, B. W., Ganesh, P., Maurya, R., Pathak, U. K., Kumar, R., Ghosh, M., & Singh, Y. (2021). Spatial distribution of outbreak of locust swarms: A geographical analysis of vulnerability and preventions in India. *Sustainability, Agri, Food and Environmental Research, 9*(1), 91–106.

Pine, K., Edwards, R., Masera, O., Schilmann, A., Marrón-Mares, A., & Riojas-Rodríguez, H. (2011). Adoption and use of improved biomass stoves in Rural Mexico. *Energy for Sustainable Development, 15*(2), 176–183. https://doi.org/10.1016/j.esd.2011.04.001

Purohit, K., & Samant, S. S. (1995). Fodder trees and shrubs of Central Himalaya. Gyanodaya Prakashan, Nainital.

Raj, A., Jhariya, M. K., Yadav, D. K., Banerjee, A., & Toppo, P. (2020). Agroforestry for climate change mitigation, natural resource management, and livelihood security. In *Climate change and agroforestry systems* (pp. 27–46). Apple Academic Press.

Rao, P. P., & Hall, A. J. (2003). Importance of crop residues in crop–livestock systems in India and farmers' perceptions of fodder quality in coarse cereals. *Field Crops Research, 84*(1–2), 189–198.

Sharma, P. K. (2005). Livelihood strategy of migrant and non-migrant small and marginal farmers: A case study conducted in Aurangabad district of Maharashtra. *Jharkhand Journal of Development and Management Studies, 3*(4), 1677–1703.

Singh, D. V. (1990). *Production and marketing of off-season vegetables*. Delhi: Mittal Pubulishers.

Surayya, T., Krishna, K., Sharma, R., Karla, S., Kujur, S. S., Bala, S., & Basnayat, B. (2008). Sericulture based micro enterprise as a source of rural livelihood and poverty alleviation: A case study of Anantapur. *Journal of Rural Development, 27*(1), 149–176.

World Bank. (2006). *The World Bank Annual Report 2006*. The World

Chapter 4
Estimation of Forests Degradation in the Tiger Corridor of Rajasthan, India: In the Context of Land Use Change and Local Sustainability

Bhanwar Vishvendra Raj Singh, Anjan Sen, Ritika Prasad, and Ravi Mishra

Abstract Global deforestation continues at an alarming rate in 2015–2020. According to UNEP, ten million hectares of forest are diverted to other land uses every year, which is roughly the size of Iceland. Forests are home to many flora and fauna species, and the ecosystem provides provisional, ancillary, regulation, and cultural services. The aim of this study is to assess forest cover changes in the Tiger Corridor of Rajasthan using satellite imagery and climate data. A supervised classification was performed on two multi-temporal satellite imageries, and a total of six major land use and land cover (LULC) classes were identified and mapped. Using post-classification techniques, it is noted that the forest cover (dense forest + shrubland) has decreased from 2673.92 to 2289.106 km^2, respectively, from 2000 to 2020. NDVI analysis was performed on these images, and it was noted that there was no significant difference between NDVI classification and supervised classification of images. SAVI and RDVI classifications have also been used to check vegetation accuracy. Ultimately, the results indicate that from 2000 to 2020, forest cover was reduced by 384.814 km^2 which is about 24.48%. The correlation coefficient of NDVI and climatic factors were also explored to examine the relationship between changes in climate and vegetation. The result shows that there is a very positive relationship between climate change and vegetation change. This outcome contributes to sustainability and regional environmental management for all stakeholders for Tiger

B. V. R. Singh (✉)
Department of Geography, Faculty of Earth Sciences, Mohanlal Sukhadia University, Udaipur, India
e-mail: bhanwarsa28@mlsu.ac

A. Sen
Department of Geography, Delhi School of Economics, University of Delhi, New Delhi, India

R. Prasad
Department of Geography, University of Lucknow, Lucknow, India

R. Mishra
Ex-Student of Department of Remote Sensing and GIS, SSJ Campus - Almora, Kumaun University, Almora, India

© The Author(s), under exclusive license to Springer Nature Switzerland AG 2024
B. V. Raj Singh and A. K. Batar (eds.), *Sustainable Local Development for Environmental and Social Sustainability*, Human-Environment Interactions 11,
https://doi.org/10.1007/978-3-031-67303-0_4

Corridor. Eventually, the research focuses on the role of forest ecosystem services in the social-ecological inclusion of Ranthambhore areas, which is possible through cultural ecosystem services, maximum public participation, development of indigenous community and as well as restoration of the wildlife corridor and sustainable planning. The result of this study will aid in the formulation of forest resource management, land use plans as well as corridor conservation actions, all of which will contribute to the region's long-term wildlife sustainability.

Keywords Forest degradation · Anthropogenic pressure · Biodiversity conservation · Sustainability

4.1 Introduction

Forest ecosystem is very rich in resources and delivers important ecosystem goods and services. Forests are home to a large number of plant and animal species. Forest is not only nature's beauty to be admired but is also a phenomenon which fulfills the basic need of humans and its society such as by providing fuel woods, bioenergy, eatable material like fruits and vegetables, medicines, and shelter to many grazers and farmers (Coates, 2013).

India is a vast country that is surrounded by many sources such as forests, settlements, oceans, rivers, reservoirs, mountain ranges, Western and Eastern Ghats along with many natural phenomena presented (Kuriakose et al., 2009). India is rich in its natural resources, it has vast forest cover areas, all four seasons of climatic conditions have maximum types of flora and fauna found in the land like Lion, Bengal tiger, white tiger, leopard, one-horned rhinos, lion-tailed macaque, and the flora of India varies from the Western Himalayas to the plains region of the country, like chirping, blue pine, silver fur, dwarf willows; the flora of the country varies from region to region due to different weather conditions also including the forest regions from temperate to the coniferous, from evergreen to the scrubs and the deciduous forests, and also from the tropical jungles to the cool temperate woods (Agarwal, 2007). Despite India being a rich resourceful country, it is facing major problems in the forest sector majorly because of climate change and changes in land use and land cover. Problems of deforestation, degradation, and loss of habitat for both flora and fauna along with the harsh climatic conditions are impacting the forests (Palo, 1999). Forest and climate are linked to one another; forest degradation is a cause of climate change as well as its effect (Altman et al., 2024). Forests act as carbon sink but with the loss of forests, and forest fires carbon is released back into the atmosphere.

Forests are very important for the survival of the ecosystem and also the biodiversity (Putz et al., 2001). India is richly endowed with biodiversity harboring an estimated 10% of the plant species and seven percent of the animal species found within the world, which is remarkable provided that the country represents only 2% of the world's landmass (Kothari, 1997). India is ranked eighth among two-hundred-one countries in global biodiversity index, 2022 and also has four of the world's biodiversity hotspots recognized by the international conservation community. In this research paper, study area is located in Rajasthan state (North-Western state of India), In the tiger corridor, Ranthambhore Tiger Reserve is hotspot of tiger habitat (Singh & Sen, 2022).

The use of remote sensing with GIS techniques in recent times has been of enormous help in monitoring the changing the pattern of forest cover. It provides a number of the foremost accurate means of measuring the extent and pattern of changes in cover conditions over a period of time. Forest cover today is altered primarily by direct human use and any the conception of world change must include the persistent influence of human action onto the land surface conditions and processes. In recent times, the application of remote sensing combined with GIS techniques has been immensely beneficial in monitoring changes in forest cover patterns (Maguire, 1991). These technologies offer some of the most precise methods for assessing the extent and nature of changes in forest cover over time. Today, forest cover is mainly altered by direct human activities, and any understanding of global change must take into account the ongoing impact of human actions on land surface conditions and processes (Balmford et al., 2002).

Mapping LULC is the best way to monitor changes and to check land use change and development. A change detection analysis was performed to determine the forest cover changes in the study area. The results will quantify the land cover change patterns within the area and demonstrate the potential of multi-temporal satellite data to map and analyze changes in land cover in spatio-temporal framework (Brandt & Rickard, 1994). This may be used as inputs to land management and policy decisions with regard to varied themes that have a link with space like urbanization, water management, deforestation, and land degradation. The aim of this research is to research and monitor the spatio-temporal LULC change patterns using multi-temporal satellite images from the period 2000–2020 within the Tiger Corridor of Rajasthan (Sowmya & Somashekar, 2010). Concurrent with this aim is to spot and quantify the major LULC classes, to detect changes using change detection techniques, and to spot the key driver(s) of change within the study area.

4.2 Study Area

Tiger Corridor of Rajasthan is found between latitudes 25,052′071″ N to 260 33′713″ N to longitudes 75,085′84.0″ E to 77,002′48.0″ E at the junction (Great boundary fault) of Aravalis and Vindhyan ranges. Study area has a complete geographic area of Ranthambhore Tiger Reserve 1411.28 km^2. with core area of 1113.36 km^2. and buffer area is 297.92 km^2 (Jhala et al., 2015). The Tiger Corridor of Rajasthan is connected to Keoladeo, Van Vihar, Ranthambhore, Sawai Mansingh, Ramgarh, Jawahar Sagar, and Dara sanctuaries falling under Karauli, Bharatpur, Sawai Madhopur, Bundi, Kota, Chittorgarh, and Jhalawar districts. It is the only most massive expanse of dry deciduous Anogeissus forest left intact in India. Ranthambore is the foremost

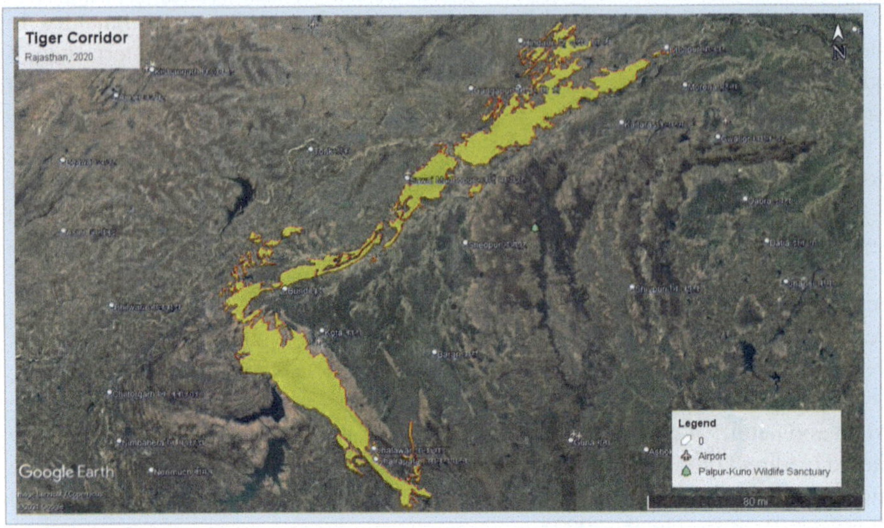

Fig. 4.1 Study site from Google Earth. *Source* Google earth

hotspot tiger reserve and marks the transition zone between the important deserts and seasonally wet peninsular India. (Fig. 4.1) (Vishvendra & Sen, 2018).

4.3 Research Methodology

The research is processed with the utilization of satellite data from various sources, and also, the multi-temporal data is taken into consideration. For data collection, Landsat 5, Landsat 8, and ASTER are used. The most important element of the research is the utilization of remote sensing and GIS (Jaiswal et al., 1999; Jensen, 2009) (Fig. 4.2).

4.3.1 Land Use and Land Cover Change Analysis

4.3.1.1 Supervised Classification

To analyze the land use and land cover change of the Tiger Corridor, Rajasthan, the supervised classification method was applied in the ERDAS Imagine software. In this method, the user specifies the various pixel value or spectral signatures, which pixel values or spectral signatures are associated with each respective class. This process is done by selecting representative sites of the known cover type which are known as "training sites or areas." In ERDAS Imagine software, these samples are collected

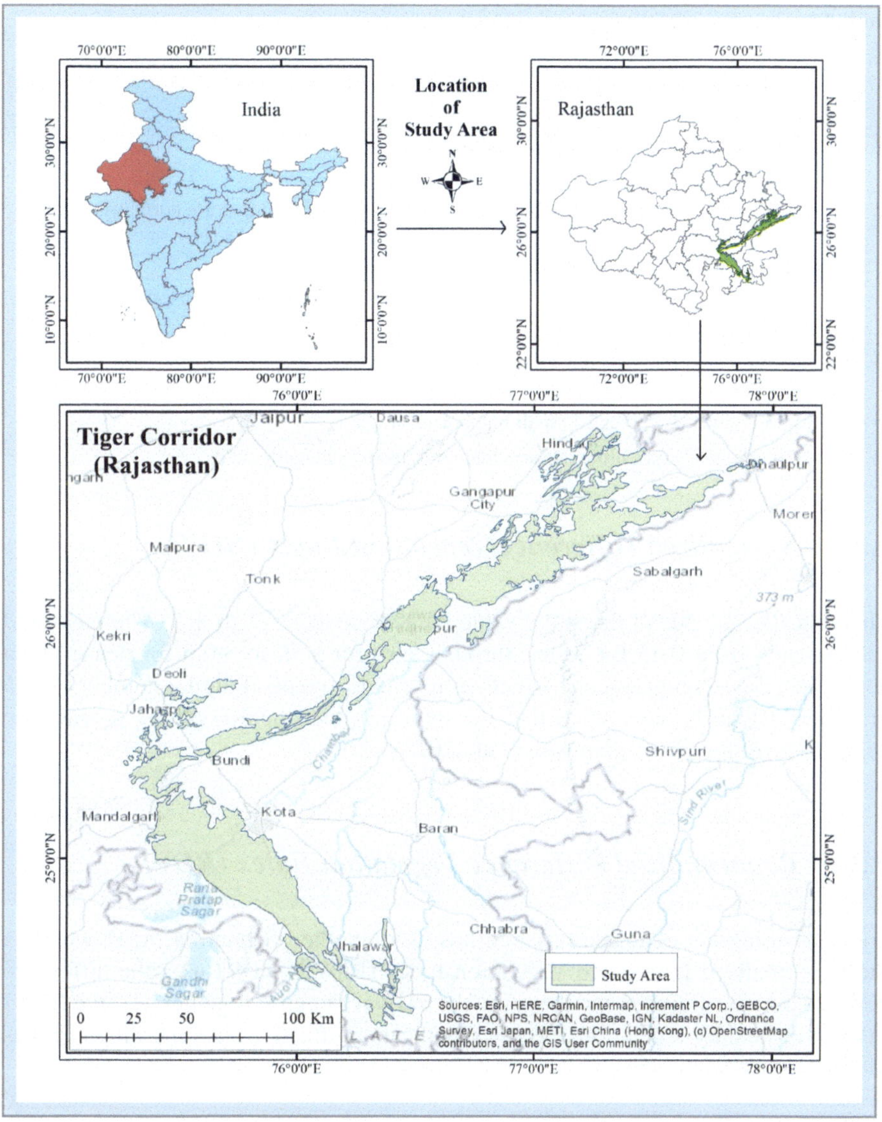

Fig. 4.2 Map of study area. *Source* Esri

from the signature editor toolbox. This classification method is used the maximum likelihood classifier algorithm (Rawat et al., 2015). This is one of the most popular supervised classification methods, which is used in remote sensing image data. This theory assumes that these probabilities are equal for all the respective classes.

The images can be classified into six respective classes; these are water bodies, fellow land, built-up land, dense forest, shrub land, and grassland (Table 4.1).

Table 4.1 Description of the land use and land cover types

S. No.	Land use/ land cover class	Description
1	Water bodies	Include rivers, ponds, streams, etc.
2	Fellow land	Includes the fellow land and all pieces of land that are normally used for farming but that are left with no crops on it for a season in order to let it recover its fertility
3	Built-up land	Includes the areas of constructions, roads, bridges, houses, etc.
4	Dense forest	Includes the areas which are covered by trees
5	Shrub land	Includes all-small plants and shrubs
6	Grassland	Agriculture shown area, grasslands, and cattle fields

4.3.2 *Normalized Difference Vegetation Index (NDVI)*

Theoretically, the NDVI threshold value ranges between -1 to $+\ 1$. The measured values range from 0.35 for water, through zero for soil, to $+0.6$ for dense green vegetation, based on grayscale, which corresponds to a pixel digital number of 135 or higher. It can be concluded that the more positive the NDVI, the greener the vegetation within a pixel (Nguyen et al., 2012).

4.3.3 *Renormalized Difference Vegetation Index (RDVI)*

For confirming the changes, research has used another vegetation index which is the Renormalized Difference Vegetation Index (RDVI). RDVI uses the difference between the near-infrared and therefore the red wavelengths, together with the NDVI to focus on healthy vegetation. It is used for the most effective outcome which applies to heavily vegetated areas.

4.3.4 *Soil-Adjusted Vegetation Index (SAVI)*

SAVI stands for soil-adjusted vegetation index which shows the irregular deviation within the inundation effect from soil moisture, soil color, and high-density vegetation. Huete (1988) established a vegetation index that accounted for different red and near-infrared obliteration concluded the vegetation canopy.

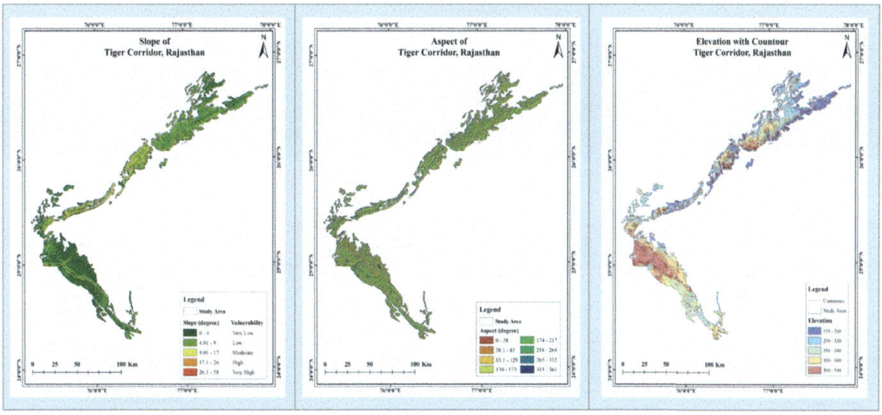

Fig. 4.3 Terrain Map of Tiger Corridor of Rajasthan. *Source* Esri

4.3.5 Terrain Mapping

Slope refers to the steepness or the degree of the incline of a surface. The slope map of the Tiger Corridor ranges from 0 being the bottom or the surface and 58 being the best slope area within the study area. In Fig. 4.3, the slope map is characterized into five classes.

Aspect is the orientation of the slope, measured clockwise in degrees from 0–360, whereas 0 is North facing, 90 is East facing, 180 is South facing, and therefore, 270 is West facing. Aspect values indicate the direction of the physical slope faces which are supported by the slope angle.

The elevation map of the Tiger Corridor is the representation of the shaded area of the slopes with the assistance of azimuth and sun angular direction. The above map represents a height value of 0, and also, the highest value is 592 m (Chart 4.1).

4.4 Result and Discussion

4.4.1 Land Use Land Cover

Land cover refers to the area or a region naturally covered by the natural environment like the forest, agriculture, water bodies, and wetlands, whereas land use refers to the development of the regions by humans for their living whether for conservation, hospitals, schools, residential, and so on for their betterment. Land cover can be defined as the analysis of both satellite and aerial imagery, and it also helps in the identification of the landscapes over a period of time.

Figure 4.4 represents the land use and land cover area of the Tiger Corridor of Rajasthan in the years 2000 and 2020. The figure represents increased dense forest

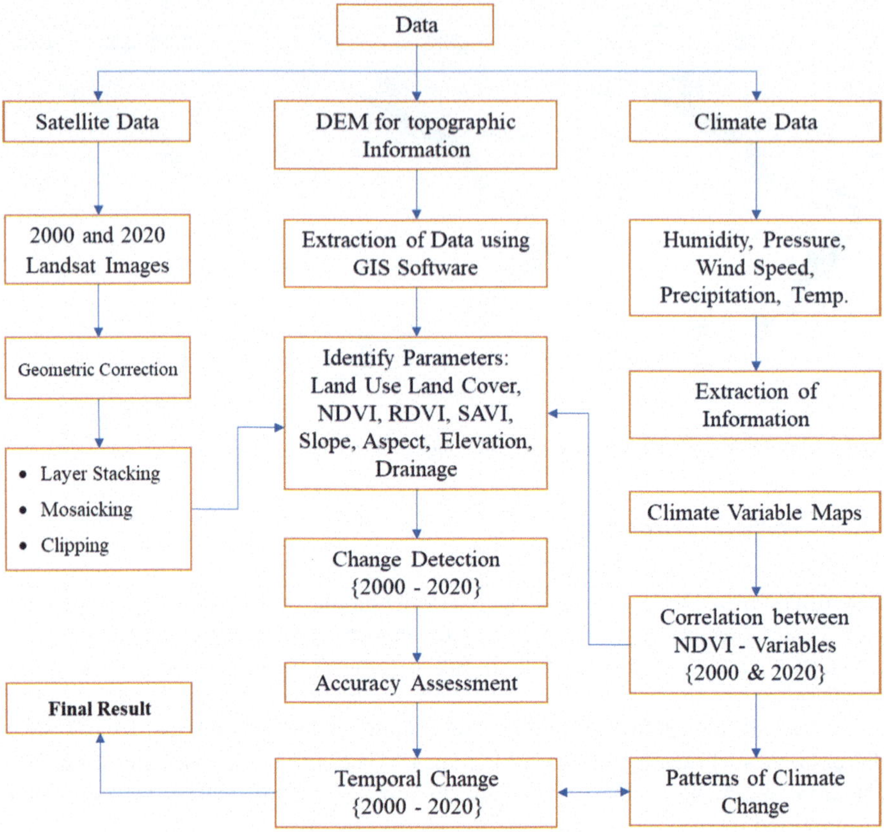

Chart 4.1. Methodology chart

area from 1290.44 km^2 in 2000 to 1930.74 km^2 in 2020. Shrub land area has decreased from the years 2000 to 2020 from 1383.48 to 358.66 km^2. Grassland consisted area of 738.77 km^2 in 2000 and increased to 3236.06 km^2 by 2020. Overall forest area along with shrub land has decreased by 384.81 km^2 in recent years and is about 24.48% (Table 4.2).

4.4.1.1 Accuracy Assessment

During the year 2000, Southern areas of the Tiger Corridor have a continuous chain of dense forests encompassed with shrubs and running from North-East to middle part of the area. The signs of no forest land are visible in the middle part and little patches in the Southern areas. The shrubs can be seen covering most of the upper areas. Dense forest has increased in recent years while shrub land has decreased; it is clearly visible in the maps as greenness is less in LULC 2020. The high vegetation

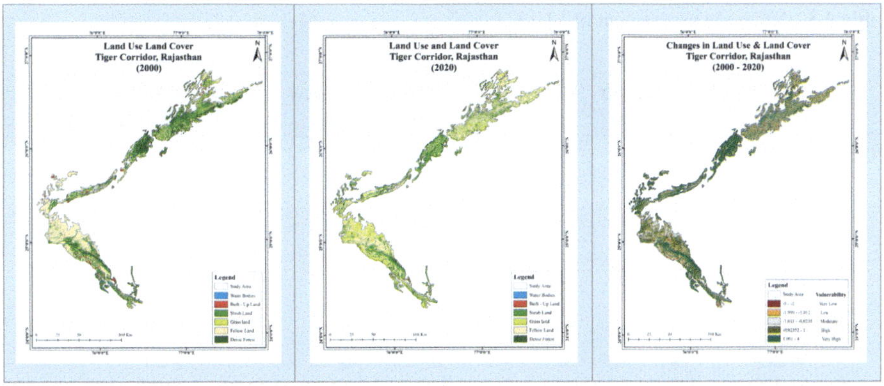

Fig. 4.4 LULC and LULCC Map of 2000 and 2020. *Source* Esri

Table 4.2 Changes in LULC between 2000 and 2020 (Ha)

S. No	LULC class	2000	2020	(2020–2000)	Change percentage
1	Water	65.82	38.4524	−27.3676	−41.57
2	Shrub land	1383.48	358.366	−1025.114	−74.09
3	Grassland	738.77	3236.06	2497.29	338.03
4	Dense forest	1290.44	1930.74	640.3	49.61
5	Fellow land	3246.19	1147.56	−2098.63	−64.64
6	Settlement	54.12	12.9991	−41.1209	−75.98

is only visible in the Ranthambore tiger reserve part of the area which falls in the middle of the area; in the upper part of the area, little bit of shrub lands is visible. But there are some increases in the vegetation in the Southern part of the area compared to the 2000 data. Grassland has increased remarkably by almost 338 % (Table 4.3).

The maximum values of the normalized difference vegetation index were reduced from 0.63 in 2000 to 0.53 in 2020 (Table 4.4). The darker the green color in the image, the highest NDVI values were recorded and the highest vegetation cover. In the same way, the minimum values also decreased from −0.37 in 2000 to −0.21 in 2020. This means that status of vegetation in 2000 is better than the vegetation in 2020. But there are some other factors also present because the lowest value is also decreasing which means the other small type of vegetation has increased in these years, e.g., grassland.

Figure 4.5 of RDVI for the year 2000 indicates the high and low values with 7.60 to −3.02, respectively. Another figure for the year 2020 indicates the high and low values with 99.11 to −29.66, respectively.

SAVI Maps are also created to analyze the vegetation change, as there are some conflicts between NDVI and RDVI as one is showing increasement while the other is showing decreasing pattern in vegetation. From Table 4.4, it is clearly noticeable that

Table 4.3 Accuracy assessment of LULC

		2000						
		Built-up	Grassland	Dense forest	Fallow land	Shrub land	Water bodies	Row total
2020	Built-up	1767	7	0	0	0	0	1774
	Grassland	0	942	0	0	0	0	942
	Dense forest	4	0	682	8	0	0	694
	Fallow land	5	5	20	146	0	0	176
	Shrubland	1	0	0	0	67	1	69
	Water bodies	0	0	0	0	3	70	73
	Column total	1777	954	702	154	70	71	**3728**
	User accuracy	**1.00**	**1.00**	**0.98**	**0.83**	**0.97**	**0.96**	
	Producer accuracy	**0.99**	**0.99**	**0.97**	**0.95**	**0.96**	**0.99**	
	Overall accuracy	**98.55**						
	Kappa coefficient	**0.98**						

These are Important part of tables therefore it's bold.

Table 4.4 Statistics of NDVI, RDVI, and SAVI of 2000 and 2020

Statistic	Minimum		Maximum		Mean		Standard deviation	
Vegetation indices	2000	2020	2000	2020	2000	2020	2000	2020
NDVI	−0.358	−0.21	0.63	0.53	0.164	0.128	0.092	0.028
RDVI	−3.023	−29.67	7.61	99.12	1.57	21.38	0.87	4.669
SAVI	−0.53	−0.316	0.95	0.79	0.245	0.19	0.138	0.042

the mean and SD values decreased in these years. Hence, the vegetation is decreased (Shen et al., 2016).

4.4.2 *Forest Mapping*

See Fig. 4.6.

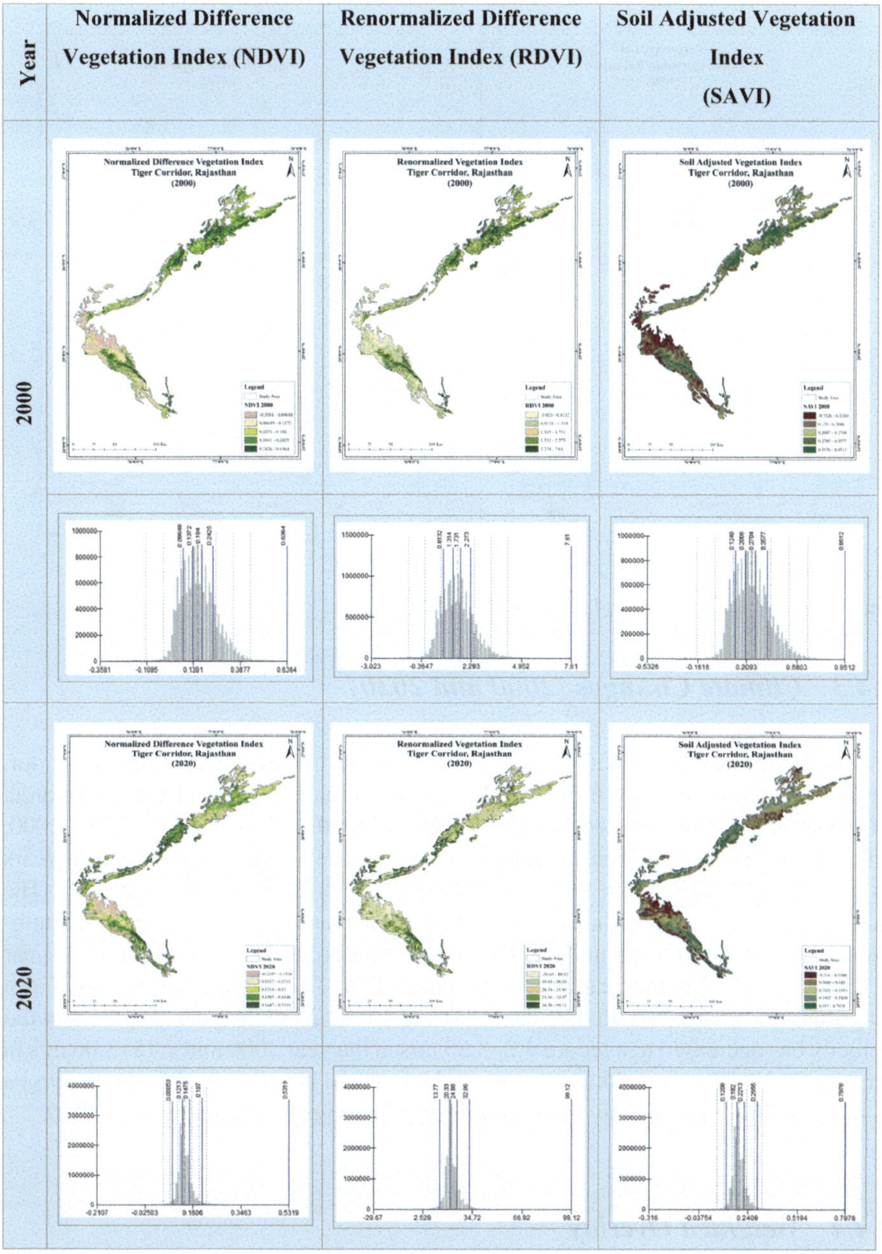

Fig. 4.5 NDVI, RDVI, and SAVI Maps of 2000 and 2020. *Source* Esri

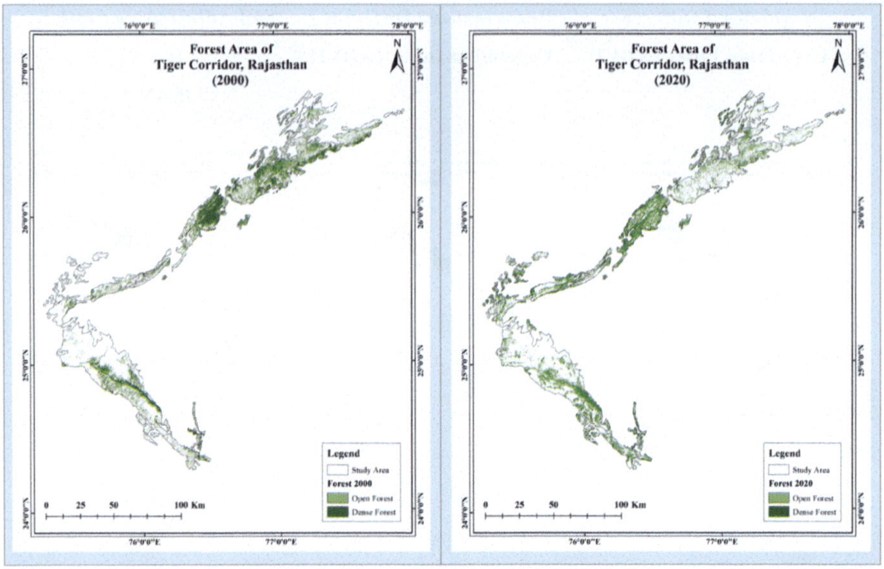

Fig. 4.6 Forest area of 2000 and 2020. *Source* Esri

4.4.3 Climate Changes (2000 and 2020)

Available meteorological data (Fig. 4.7) represent the climate data of Tiger Corridor, Rajasthan. These data were collected from 14 climate centers in the study area. Climatic information is based on annual averages for the 20-year period 2000–2000, reflecting the following: The annual mean temperature has decreased in recent years as it was 26.29–27.4 °C in the year 2000 and 25.36–26.38 in the year 2020. The rainfall was very variable; the amount of rainfall increased from North-East to South-West in both years. The annual rainfall average was around 0.82–1.6 mm in the year 2000 and 1.3–1.71 in the year 2020. The annual humidity average has increased as it was around 31.5–35.31% in the year 2000 and 46–52.28 in the year 2020. The wind velocity has decreased to a record 3.2–3.55 m/s in the year 2000 and 2.75–3.06 m/s in the year 2020. The annual mean pressure data ranged between 96.44 and 98.59 kPa in the year 2000 and slightly increased to 96.57 and 98.77 kPa in the year 2020.

4.4.4 Weighted Overlay

See Fig. 4.8.

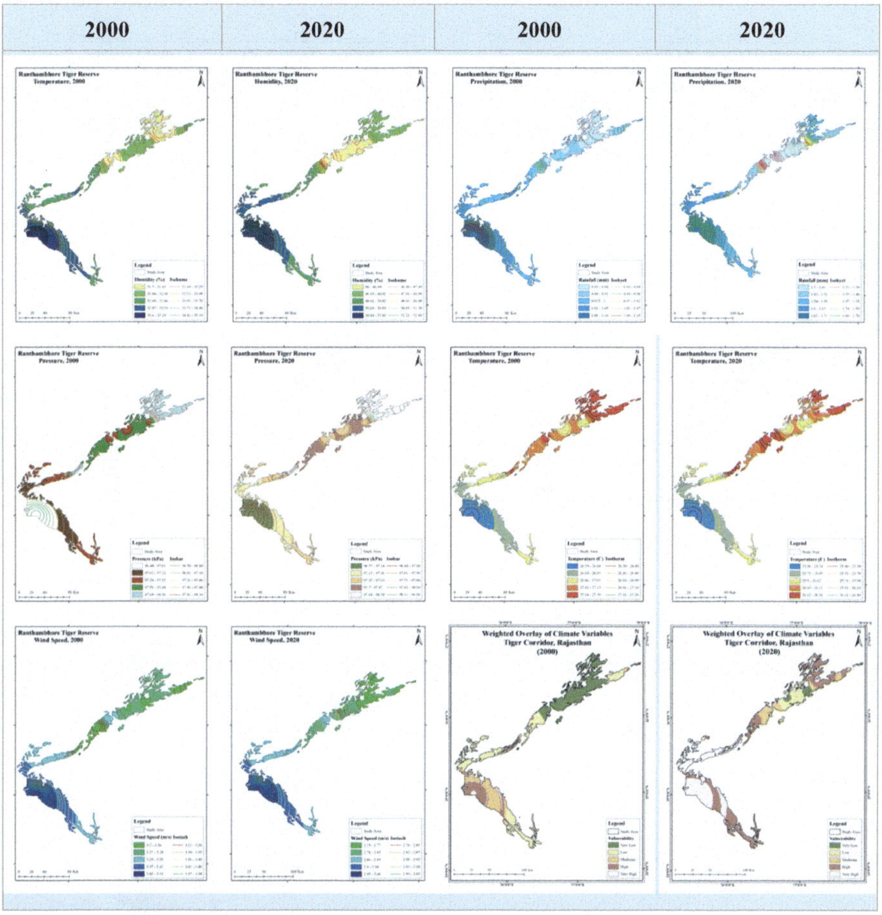

Fig. 4.7 Climate variables map of 2000 and 2020. *Source* Esri

4.4.4.1 Final Weighted Overlay

In this study, forest-ecological areas are considered to be land bodies that are uniform in terms of topography, vegetation, and climate for the purposes of planning. They are unique in that specific combination of landforms, vegetation, and climate is used, which are considered inputs for the mapping of forest-ecological zones.

4.4.5 Correlation Between NDVI and Climate Variables

The results showed that NDVI in the year 2000 has a positive relationship with temperature and pressure, but on the other hand, it is negative with other variables

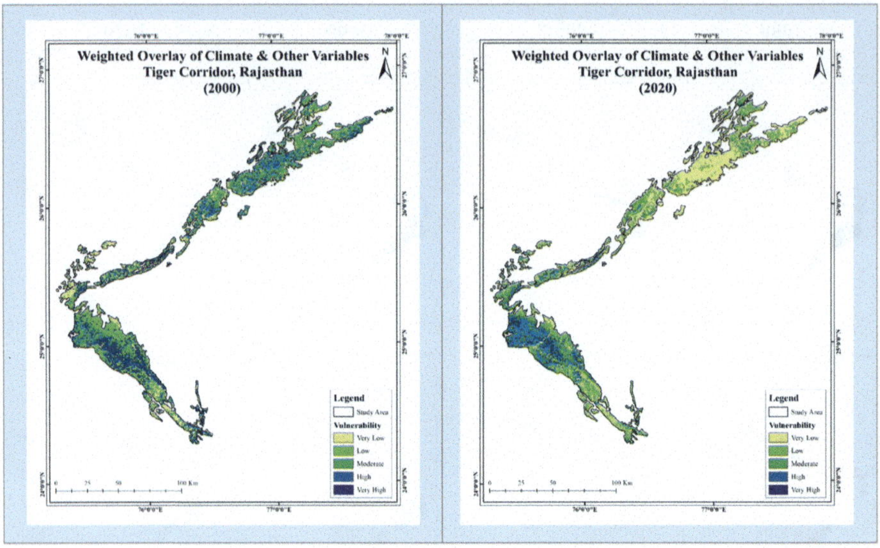

Fig. 4.8 Weighted overlay Maps for years 2000 and 2020. *Source* Computed by the researcher

such as, humidity, precipitation, and wind speed. The result has been changed in the year 2020 as this time there is only a negative correlation with temperature while other variables have a positive correlation with NDVI (Fig. 4.9).

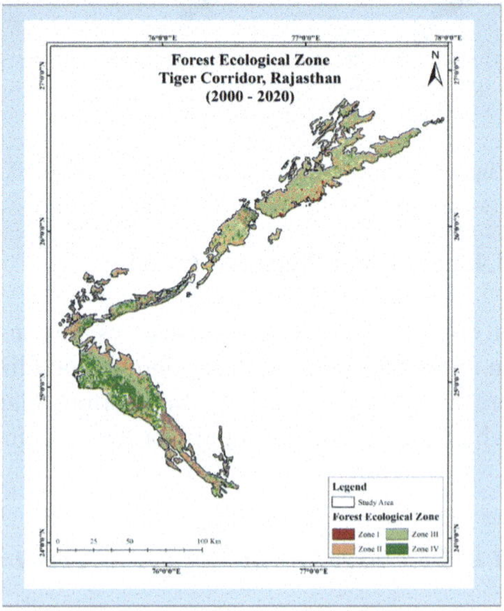

Fig. 4.9 **Final** weighted overlay for years 2000 and 2020. *Source* Computed by the researcher

4.4.6 Ground Truth Surveys

In order to improve the completeness and utility of the data collection, resulting in better decision making and a better comprehension of different environmental and geographic phenomena, ground survey has been carried out in the area under study.

4.5 Discussion

This study is assessing and analyzing the relationship between NDVI and climate variables in the Ranthambhore Tiger Reserve (RTR) (2000–2020). The relationship between climatic factors and NDVI values of RTR from 2000 and 2020 was analyzed by simple linear regression and correlation coefficients. In addition, the land cover data of the (Landsat) are used to assess land use change from 2000 and 2020 and also applied to extract NDVI, RDVI, and SAVI.

Vegetation growth is mainly driven by hydrothermal conditions. Due to the monsoon climate in the study area, rainfall and temperature are coordinated in summer, and high temperatures and abundance of summer rainfall generally correspond to a high vegetation index.

There are several reasons for the high rate of grassland change observed in the study area. Some of the reasons include changes in the weather patterns such as more rainfall or more favorable monsoon seasons. Since Rajasthan is semi-arid, any increase in rainfall can encourage grass and other vegetation growth. If there is a concerted effort to preserve and restore degraded land or to protect natural habitats, this could lead to an increase in the number of grassland areas. The health and productivity of the grasslands can be improved through better land management practices such as sustainable grazing or rotational grazing or controlled burning.

Agricultural practices such as converting cultivated land back into grassland or fallow land could lead to an increase in the number of grassland areas. Sometimes, abandoned or unused farmland may undergo natural succession in which grasses and other vegetation gradually colonize the land, resulting in grasslands. Grasslands can host and support a wide range of wildlife species, which can further encourage the growth and upkeep of grasslands.

Like many other regions around the world, Rajasthan has been striving to meet the UN SDGs, which provide a framework to countries and communities to tackle various social, economic, and environmental challenges (Canadell & Raupach, 2008; Chivian, 2002). The degree to which people in the state are on track to meet the SDGs varies from sector to sector and community to community. Some areas where progress has been made, or where efforts are being made, include education, poverty alleviation, healthcare, clean energy and climate action, gender equality, water and sanitation, sustainable agriculture, innovation, entrepreneurship, and sustainable industrialization.

4.6 Conclusion

The relationship between forest cover and their associated LULC classes was investigated, and various thematic maps were developed. The most LULC types identified within the catchment area are companion land, built-up, grassland, dense forest, water body, and shrubland. It was observed that there has been a significant change in vegetation from the 2000–2020 volume, where there has been an increase in the dense forest cover and a decrease in the shrubland. But on the other hand, there was also a significant increase in grassland. There is generally both dense forest and shrubland are included in vegetation cover, so the result has been shown to be a reduction in overall vegetation cover.

To analyze the climate effect on vegetation, five climate variables have been selected, viz., humidity, mean temperature, wind speed, precipitation, and pressure for correlation with NDVI. The changes in vegetation may be a result of climate change. The vegetation in the study areas is slightly sensitive to climatic factors. Simple linear regression is also applicable to NDVI and climatic factors (Fig. 4.10). In general, the NDVI values decrease at very high temperatures and at very low temperatures (Fig. 4.11).

Anthropogenic activities within the study area can be other factors for vegetation changes, but these were not investigated. Therefore, further analysis is needed to higher explain the impact of these factors on forest cover change. Future studies are recommended to assess the prediction of forest cover, which is in danger of invasion.

According to forest department officials, the realm from Kota to Dholpur was considered a corridor for tigers in the past too. The animals are accustomed to roaming freely between these areas; however, deforestation has changed matters in recent times. Ranthambore has witnessed a rise in the number of tigers in recent years due to strong conservation efforts; hence, the necessity to create and monitor the different corridors has become necessary. This may also make sure that human encroachment is avoided. Achieving the sustainable development goals (SDGs) is a shared responsibility that involves government, civil society, business, and individuals. Progress toward the SDGs can be affected by a variety of factors, such as government policies, funding, technological progress, awareness, and societal engagement. Human beings can play an important role in increasing and maintaining natural resources through responsible, sustainable practices. The efforts of individuals can significantly increase and maintain natural resources.

It is essential that people come together, take ownership, and work for a sustainable future that will benefit current and future generations.

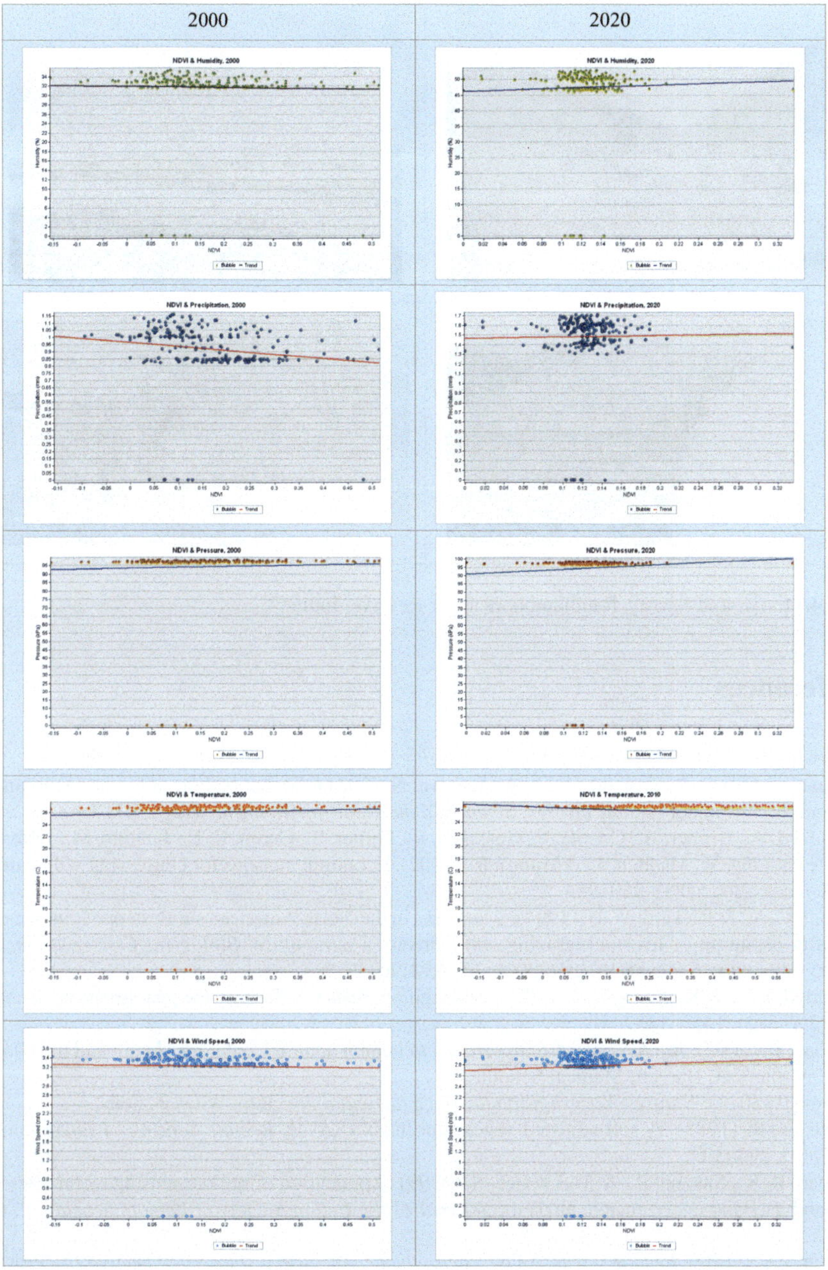

Fig. 4.10 Climate and NDVI Change 2000 and 2020. *Source* Computed by the researcher

Fig. 4.11 Ground Survey, Ranthambhore Tiger Reserve, India

References

Agarwal, B. (2007). *Environmental issues in India: A reader*. Pearson Education India.

Altman, J., Fibich, P., Trotsiuk, V., & Altmanova, N. (2024). Global pattern of forest disturbances and its shift under climate change. *Science of the Total Environment, 915*, 170117.

Balmford, A., Bruner, A., Cooper, P., Costanza, R., Farber, S., Green, R. E., Jenkins, M., Jefferiss, P., Jessamy, V., Madden, J., &Munro, K. (2002). Economic reasons for conserving wild nature. *Science, 297*(5583), 950–953.

Brandt, C. A., & Rickard, W. H. (1994). Alien taxa in the North American shrub steppe four decades after cessation of livestock grazing and cultivation agriculture. *Biological Conservation, 68*, 95–106. https://doi.org/10.1016/0006-3207(94)90339-5

Canadell, J. G., & Raupach, M. R. (2008). Managing forests for climate change mitigation. *Science, 320*(5882), 1456–1457.

Chivian, E. (2002). *Biodiversity: Its importance to human health.* Center for Health and the Global Environment, Harvard Medical School.

Coates, P. (2013). *Nature: Western attitudes since ancient times.* John Wiley & Sons.

Huete, A. R. (1988). A soil-adjusted vegetation index (SAVI). *Remote Sensing of Environment, 25*(3), 295–309.

Jaiswal, R. K., Saxena, R., & Mukherjee, S. (1999). Application of remote sensing technology for land use/land cover change analysis. *Journal of the Indian Society of Remote Sensing, 27*(2), 123.

Jensen, J. R. (2009). *Remote sensing of the environment: An earth resource perspective* 2/e. Pearson Education India.

Jhala, Y. V., Qureshi, Q., & Gopal, R. (Eds.). (2015). The status of tigers, copredators & prey in India 2014. *National Tiger Conservation Authority, New Delhi & Wildlife Institute of India, Dehradun.* TR2015/021

Kothari, A. (1997). *Conserving India's agro-biodiversity: Prospects and policy implications. Gatekeeper Series no. 65,* International Institute of Environment and Development.

Kuriakose, S. L., Sankar, G., & Muraleedharan, C. (2009). History of landslide susceptibility and a chorology of landslide-prone areas in the Western Ghats of Kerala, India. *Environmental Geology, 57*(7), 1553–1568.

Maguire, D. J. (1991). An overview and definition of GIS. *Geographical Information Systems: Principles and Applications, 1*, 9–20.

Nguyen, T. T. H., De Bie, C. A. J. M., Ali, A., Smaling, E. M. A., & Chu, T. H. (2012). Mapping the irrigated rice cropping patterns of the Mekong delta, Vietnam, through hyper-temporal SPOT NDVI image analysis. *International Journal of Remote Sensing, 33*(2), 415–434.

Palo, M. (1999). *No end to deforestation? In World forests, society and environment* (pp. 6577). Springer.

Putz, F. E., Blate, G. M., Redford, K. H., Fimbel, R., & Robinson, J. (2001). Tropical forest management and conservation of biodiversity: An overview. *Conservation Biology, 15*(1), 7–20.

Rawat, S., Gupta, A. K., Sangode, S. J., Srivastava, P., & Nainwal, H. C. (2015). Late Pleistocene–Holocene vegetation and Indian summer monsoon record from the Lahaul, northwest Himalaya, India. *Quaternary Science Reviews, 114*, 167–181.

Shen, B., Fang, S., & Yu, W. (2016). Different correlations between NDVI and meteorological factors at temporal-time scales. *J. Remote Sens., 20*, 481–490.

Singh, B. V. R., & Sen, A. (2022). Assessment of Tiger Tourism in India: A case study of Ranthambore Tiger Reserve, Rajasthan, India. In R. B. Singh, M. Kumar, & D. K. Tripathi (Eds.), *Remote Sensing and Geographic Information Systems for Policy Decision Support*. Advances in Geographical and Environmental Sciences. Springer, Singapore. https://doi.org/10.1007/978-981-16-7731-1_15

Sowmya, S. V., & Somashekar, R. K. (2010). Application of remote sensing and geographical information system in mapping forest fire risk zone at Bhadra wildlife sanctuary, India. *Journal of Environmental Biology, 31*(6), 969.

Vishvendra, R. B., & Sen, A. (2018). Geo-spatial mapping of land use and land cover changes in the core and periphery area of Ranthambore Tiger Reserve, Rajasthan, India, 1975–2015. Annals of Valahia University of Targoviste. *Geographical Series, 18*(1), 62–67.

Chapter 5
Analysing Land Use and Land Cover Change for Sustainable Coastal Management: A Case Study of Southern Coastal Gujarat, India

Ritika Prasad and Bhanwar Vishvendra Raj Singh

Abstract Coastal area supports local economics based on fisheries and agriculture, apart from offering diverse opportunities for recreation and tourism. For these activities to sustain in their natural environment, protection of coastal land is the need of hour. Understanding the land use pattern through change detection of coastal areas is critical for coastal resource management, coastal environmental protection and sustainable coastal development. The current research paper focuses on the coastline of southern Gujarat, which includes districts of Bahruch, Surat, Navsari and Valsad. The focus of the study is to comprehend the dynamics of land use pattern by analysing the land use and land cover changes from 1980 to 2020 using remote sensing and GIS techniques. Assessment of land use and land cover (LULC) by remote sensing are critical to improve understanding of social, economic and ecological processes that lead to land use change in the study area. We have used accuracy assessment method and change matrix detection method which have not been taken up by the previous studies for the study area. The study suggests that changes in the coastal landscape have occurred over the decades. Results have shown that there has been an overall increase in the plantation cover and built-up area of 12.9% and 10.79%, respectively. There is significant decrease in the agriculture area −22.41%. Spatial and temporal changes in LULC along the coasts have provided the first step for identifying changes in this large area. Appropriate coastal land use planning to attain a sustainable future is important in this coastal zone, and therefore, it provides a platform to upcoming researchers for achieving conservation actions.

Keywords LULC · Planning · Sustainability · Coastal management

R. Prasad (✉)
Department of Geography, Faculty of Arts, Lucknow University, Uttar Pradesh, Lucknow 226007, India
e-mail: prasad.ritika24@gmail.com

B. V. R. Singh
Department of Geography, Faculty of Earth Sciences, Mohanlal Sukhadia University, Udaipur, Rajasthan 313001, India

© The Author(s), under exclusive license to Springer Nature Switzerland AG 2024
B. V. Raj Singh and A. K. Batar (eds.), *Sustainable Local Development for Environmental and Social Sustainability*, Human-Environment Interactions 11,
https://doi.org/10.1007/978-3-031-67303-0_5

5.1 Introduction

Land is a very essential natural resource, which includes soil, water and related vegetation and wildlife comprising the entire ecosystem. It is a fundamental resource which is limited for most human activities including cultivation, forestry, manufacturing, settlements and recreational activities, and therefore, it has been associated with economic growth. But with increasing technology and rapid industrialisation, it creates stress on the land system. The usage of land is vastly dynamic and goes through significant changes according to the varying socio-economic pattern and natural environment (Raju & Kumar, 2006).

Land use and land cover are two distinct terms which are often used interchangeably (Dimyati et al., 1996). Land use refers to how a portion of land is utilised, while land cover describes the materials that are present on the surface (Sabins, 1987). Changes in land cover by land use do not necessarily imply degradation of the land. There are many land use patterns driven by a variety of reasons and result in land cover changes that affect diversity among plants and animals, water and radiation budgets, gas emissions and other processes (Riebsame et al., 1994). Along with these the problems of sustainability of socio-economic development, it can also be closely linked with land use and land cover changes. The land use and land cover changes through its interaction with ecosystem processes, biodiversity climate and human activities drive key global changes.

In recent years, the changes in coastal land use and land cover due to human intervention have added complexities in the earth's biosphere (Jaiswal et al., 1999; Luong, 1993; Misra et al., 2013). Such complex changes are adding to the vulnerability of places and people to climatic conditions and socio, economic and political unrests of the region. (Kaliraj et al., 2015; Nicholls & Small, 2002; Yagoub & Kolan, 2006). The land use and land cover in the coastal areas are undergoing substantial transformation due to waves, wind, tide, saltwater, saltation, sea level rise, storm surge, cyclones and human interference activities (Meyer & Turner, 1994; Murali & Kumar, 2015; Muttitanon & Tripathi, 2005; Nemani & Running, 1995; Li et al., 2010; Weismiller & Momin, 1977). These are the factors which drive researchers to extensively focus on land use and land cover changes.

The significance of landscape change, both temporally and spatially, has been mentioned in a number of studies. It is vital to recognise the causes and effects of such changes on the environment at both micro- and macro levels. It is needed for the studies to elucidate the causes and effects along with recommendations on integrating the findings into plans and policies under a sustainability goal (Ayad, 2004; Baby, 2015; Brown et al., 1999; Benoit & Lambin, 2000;).

In this study, the hypothesis is set that the land use and land cover changes in the south Gujarat coastal area has undergone tremendous changes due to natural factors like waves, storm surge, cyclone and erosion and urbanisation as the main driving force. Increase in population and climatic variations has led to pressures on the land use and land cover resulting in the greatest environmental impact on vegetation cover, coastline change, landform degradation, loss of biodiversity, seawater intrusion and groundwater pollution, deterioration of soils and air along the coastal regions (Chandrasekar et al., 2001; Chauhan & Nayak, 2005; INCOIS, 2009; Mahapatra et al., 2013; Kaliraj et al., 2015). Characteristics of the change of land use patterns to other patterns are important in terms of environmental sustainability. During the last twenty years, changes in the land cover along the coastal area of south Gujarat has dramatically increased.

Analysing the landscape changes in the southern coastal Gujarat is of prime concern because coastal planning and management in Gujarat are constrained by a dearth of data about interaction between development activities and coastal environment conditions. Information relating to causes of land use change will allow more cautious preparation and organisation in the area. It will provide us with an understanding about cause and effect relationships between interacting social, economic and environment components.

This paper presents an evaluation of the change of land use as a result of natural and anthropogenic factors. Land use changes are factored from both spatial and temporal aspects. The nature of changes will provide measures to preserve the environmental capability according of each land use potential. Hence, it is crucial for procuring reliable information to support proper coastal resource management planning (Butt et al., 2015; Wu et al., 2002; Weismiller & Momin, 1977; Zoran, 2006; Zhang & Zhu, 2011).

5.2 Study Area Description

The coastline of about 417 km in South Gujarat forms a part of four districts of Gujarat such as Bharuch, Surat, Navsari and Valsad and has been investigated in the present study (Fig. 5.1).

The geographical area extends from 20°49′ N to 21°41′ N in latitudes and 73°01′ E to 73°3′ E in longitudes. The coastline of the study area is comparatively uniform and broken by few indentions. Sandy beaches are present along most of the coast of study area. Mudflats, marsh and mangrove vegetation are also found along the coast of the study area.

The Bharuch District has a population of 1,550,822 in the southern part of the Gujarat peninsula on the west coast of state of Gujarat. The Narmada River flows into the Gulf of Khambhat through its lands. The district of Surat with a total population of 6,079,231 covers an area of 7657 km^2 and is bordered by the districts of Bharuch and Narmada in the North, Navsari and Dangs in the South and the Gulf of Khambhat in the West. The town of Hazira is located in this district, which is an important trans-shipment port and is popularly known as the ‘industrial hub’ of India due to the presence of major industrial facilities. The district of Navsari covers an area of 2211 km^2 with a population of 1,329,672 and is bordered by the districts of Surat in the North and Dangs in the East. The district of Valsad is surrounded by the district of Navsari in the North, Dang in the East, and Maharashtra in the South. The population

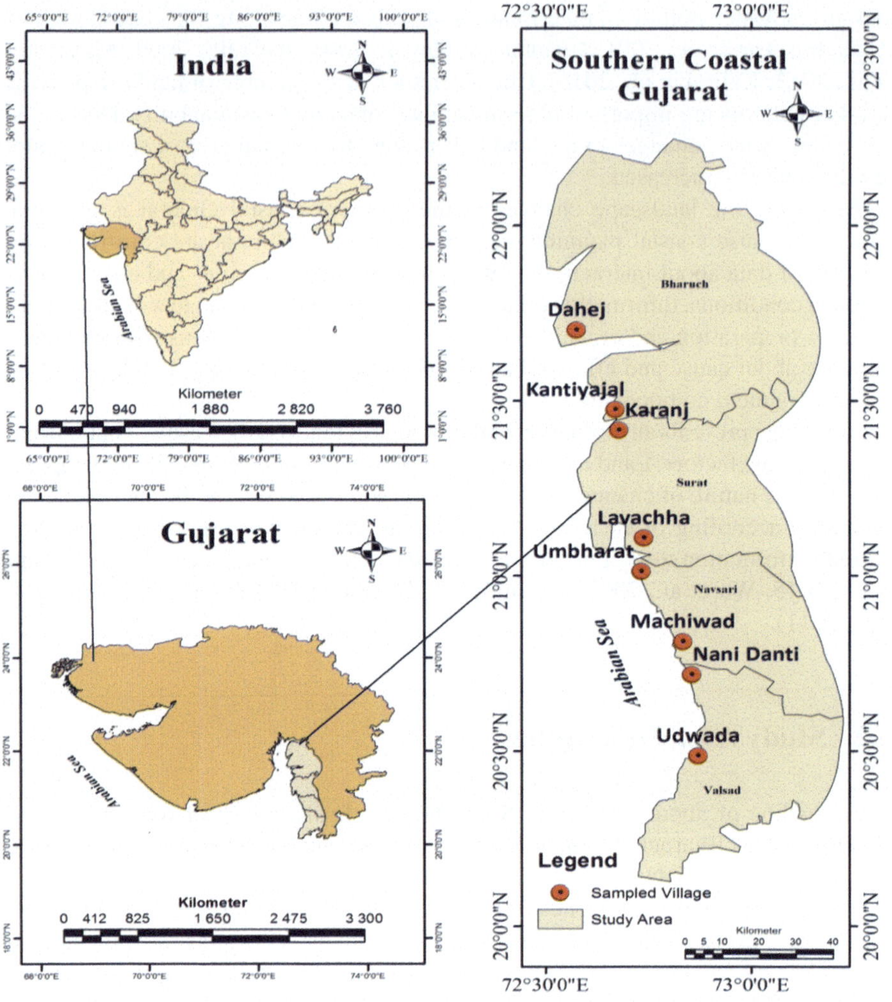

Fig. 5.1 Location map

of Valsad is 1,705,678 persons for an area of 3034 km^2 (District Census Handbook, Gujarat, 2011).

5.3 Data Sources

The present study uses multi-temporal satellite images of Landsat 3, Landsat 5 and Landsat 8 to map the changing pattern of LULC of southern coastal Gujarat for 1980, 2000 and 2020 (Table 5.1). Thematic mapper (TM), Multispectral Scanner

Table 5.1 Description of satellite images used in the study

Data	Date	Sensor	Bands used	Source
Landsat 3	23.10.1980	MSS	Band 4, Band 5, Band 6, Band 7	USGS earth explorer
Landsat 5	23.10.2000	TM	Band 1, Band 2, Band 3, Band 4, Band 7	USGS earth explorer
Landsat 8	23.03.2020	OLI	Band 2, Band 3, Band 4, Band 5, Band 6, Band 7	USGS earth explorer

System (MSS) and Operational Land Imager (OLI) sensors have been used. The main application of these sensors TM, MSS and OLI is in the areas of forest, agriculture, coastal, inland water resources and LULC mapping and monitoring. All the images were downloaded from USGS EarthExplorer website (http://earthexplorer.usgs.gov/).

Today, earth resource satellites data are useful for land use and land cover change detection studies (Yuan et al., 2005). Currently, integration of Geographical Information System (GIS) and remote sensing (RS) is one of the most important methods for detecting LULC's change. Complete methodological framework and data analysis are presented in Fig. 5.2.

5.3.1 Data Analysis

- Pre-processing

The images were layered and put on, and the study area was clipped using subset tool after converting the shapefile into area of interest (AOI). Radiometric correction was done to get more clarity in the image using histogram equalisation.

- Land cover classification scheme

To prepare the LULC map from satellite imageries, a classification scheme which outlines the LULC classes was considered. The numbers of LULC classes are preferred based on the requirement of a specific project for a particular application (Arora & Mathur, 2001; Saha et al., 2005). Six major LULC classes were chosen for mapping the entire study area consisting of vegetation; water bodies; plantations; sand bars; built-up and agriculture (Table 5.2).

- Post-processing

Fundamentally, there are two image classification techniques which are used for land cover classification. Supervised classification is one of the methods of land cover type classification using the sample polygons (ground truth points) from the known land cover types. The other type of classification is unsupervised classification, which is a

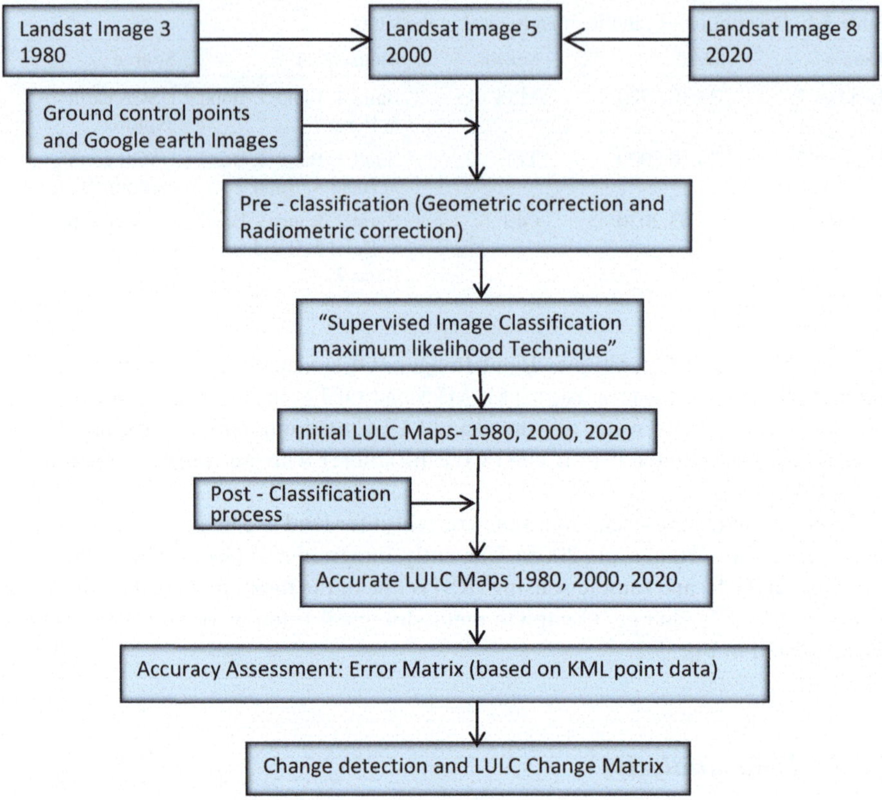

Fig. 5.2 Flowchart of methodology

Table 5.2 Description of different LULC categories

Different LULC categories	
LULC category	Classes induced general description
Vegetation	Land with tree, scrubs and area under agroforestry
Water bodies	This category comprises areas with river surface water in the form of ponds, lakes, drains and canals
Plantation	This is categorised as scattered plants and protected forest
Sand bars	Sand deposits along the coast
Built-up	This category includes urban and rural settlements, transportation, communication and recreational utilities
Agriculture	This category involves land under crops, fallow and aquaculture/ pisciculture

type of land cover classification from the satellite image data when the User does not know how many land cover types are there in the field. For this study, supervised type of classification was used using the known ground truth points and by connecting ERDAS Imagine with Google Earth and synchronising the images. In supervised classification, the classification process has been controlled by creating, managing, evaluating and editing signatures using the Signature Editor. Signatures are specific areas to which the names are assigned for supervised classification. Signatures are used to break the different classes like forest, cultivated, water and the like classes into as many subclasses as per classification requirement. After the preparation of classification scheme one of the most widely used image classification technique, that is, maximum likelihood classification was applied for mapping all the land use and land cover classes. After classifying the image using supervised classification, the accuracy of classification was checked using ground control points.

- LULC change detection

For LULC maps, post-classification change detection technique was used for analysing the changes. In the last few years, many change detection methods have been developed such as image differencing, post-classification change matrix, comparison technique and principal component analysis (Lu et al., 2004). Adopting change detection matrix method presents a vital information about the spatial distribution of changes in LULC (Shalaby & Tateishi, 2007). Change matrix showing the land cover changes in each decade was generated from classified images of 1980 to 2000, 2000 to 2020, and a change matrix was generated from 1980 to 2020 to assess the overall changes in LULC classes (Chander et al., 2009).

Accuracy assessment is important and useful in change detection. Usually, classification accuracy is done by comparing of two data sets. One is based on reference information, denoted as 'ground truth', and the other is the outcome of the analysis of remotely sensed data (Fenglei et al., 2008). Error matrix, one of the most commonly used accuracy assessment techniques was selected to assess the accuracy of classified map of 1980, 2000 and 2020. Accuracy assessment was done using a high resolution satellite imagery. Having identified data set, the next step in accuracy assessment is to generate an error matrix. Normally, the first step in creating an error matrix is to overlay the reference data on the remotely sensed classification. Since the location of each reference data set point or polygon is known, it is possible to extract the equivalent pixel values on the remotely sensed land cover image. An error matrix is a table that has the land cover classes in the reference data set as columns and the land cover classes in the remotely sensed data set as rows. The individual cells within the table indicate the area or number of pixels with a given combination of reference data and remotely sensed land cover classes. Since the two sets of classes should be identical, the diagonal line shows the pixels that have the same land cover class in both data sets. Off-diagonals have been misclassified, which indicates the error between reference data and remotely sensed data (Fan et al., 2008).

There are two categories of errors, that is, omission and commission error that take place during the classification process. Errors of commission take place when a classification process allocates pixels to a particular class that does not belong to it.

The number of pixels that are erroneously allocated to a class occurs in column cells of the class above and below the main diagonal. The Producer's accuracy defines the number of errors of commission. For each class, errors of omission take place when pixels that have its place in one class are incorporated into other classes. The user's accuracy is one more parameter that characterises the errors of omission. Furthermore, the nonparametric Kappa test was implemented to measure the level of the classification accuracy to account for diagonal elements (Table 5.7).

5.4 Results

5.4.1 LULC Status

Multi-temporal LULC covering six major classes: agricultural land, vegetation, built-up area, sand bars, plantation and water bodies of 1980, 2000 and 2020 are shown in Table 5.2. The spatial distribution pattern of LULC obtained from supervised classification is registered in Table 5.3.

Figure 5.3a–c shows the spatial distribution of the major land use and land cover classes in the study area for the years 1980, 2000 and 2020. The areal extent of these LULC features in km^2 and area coverage in per cent is shown in Table 5.3. Results from classified maps indicate that in 1980 area was occupied by different classes; agricultural land was about 27.08%, plantation was 12.61%, built-up area covered 7.4%, sand bars covered 5.93%, water bodies occupied 1.15% of the area and vegetation covered most part of the study area and occupied about 47.42%. On the other hand, in 2000 about 15.93% of the area was covered by agricultural land against 27.08% area in 1988, this showed a decrease in cultivated land. Plantation and built-up area covered 13.52% and 14.29%, respectively, while vegetation had the largest share of 47.91%. The area covered by water bodies was 1.13% of the total geographical

Table 5.3 LULC distribution in southern coastal Gujarat

LULC distribution						
Classes	Area in km^2 in 1980	Percent of area in 1980	Area in km^2 2000	Percent of area in 2000	Area in km^2 2020	Percent of area in 2020
Vegetation	65.52	47.42	63.59	47.91	62.29	45.07
Water bodies	1.58	1.15	1.48	1.13	270	1.95
Plantation	17.21	12.61	18.02	13.52	26.69	19.34
Sand bars	5.93	4.34	9.6	7.22	1020	7.38
Built-up	10.21	7.4	18.96	14.29	24.86	17.98
Agriculture	35.94	27.08	21.13	15.93	11.45	8.28
Total	136.39	100	132.78	100	138.19	100

area. Results from the classified image of 2020 illustrate that 45.07% of the area was covered by vegetation, whereas plantation share was 19.34%. Agricultural land occupied 8.28%; built-up area covered 17.98% while sand bars and water bodies covered 7.38 and 1.95% area, respectively (Fig. 5.4).

5.4.2 Accuracy Assessment

The overall LULC classification levels for the three years (1980, 2000 and 2020) ranged from 83.50% to 88.90%. The accuracy assessment revealed that overall accuracy was close to 85% for 2000 and 2020 and 88.90% for 1980 which is in proximate to Anderson's standard of 85%. The overall Kappa statistics was 0.761, 0.795 and 0.799 for 1980, 2000 and 2020, respectively (Anderson et al., 1976). Accuracies for individual LULC class are shown in Tables 5.4, 5.5 and 5.6, and Producer and User accuracy of all the classes are presented in Table 5.7. The results from the matrix provide a key platform for the following analysis of LULC changes.

5.4.3 LULC Changes

Numerous studies analyse changes in broad LULC in the coastal region yet information and data on changes in spatial–temporal patterns of LULC, driving factors and their implications within the context of sustainable development is inadequate. The LULC change matrix in the periods 1980–2000, 2000–2020 and 1980–2020 are shown in Tables 5.8, 5.9 and 5.10 respectively. The values shown in bold letters indicate no change in LULC categories for the given time period.

5.5 Discussions

5.5.1 LULC Patterns Over the Period of 1980–2000

For the period 1980 to 2000 major changes were observed from agricultural land to vegetation land. About 11.20 km^2 of agricultural land got converted to vegetation land. As the sea level is rising, it is causing the salt water to intrude on the inner parts of the coastal land. The intrusion of salt water is turning the fertile land unproductive for agriculture. Instead to keep the land barren, planting of salt-resistant trees and scrubs is done to prevent erosion from sea and winds. This can protect the coasts from wind and water erosion and storm surges. Considering the built-up area, about 5.2 km^2 of agricultural land was converted into the built-up area. The built-up areas are directly proportional to the growth of population, and it increases predominantly

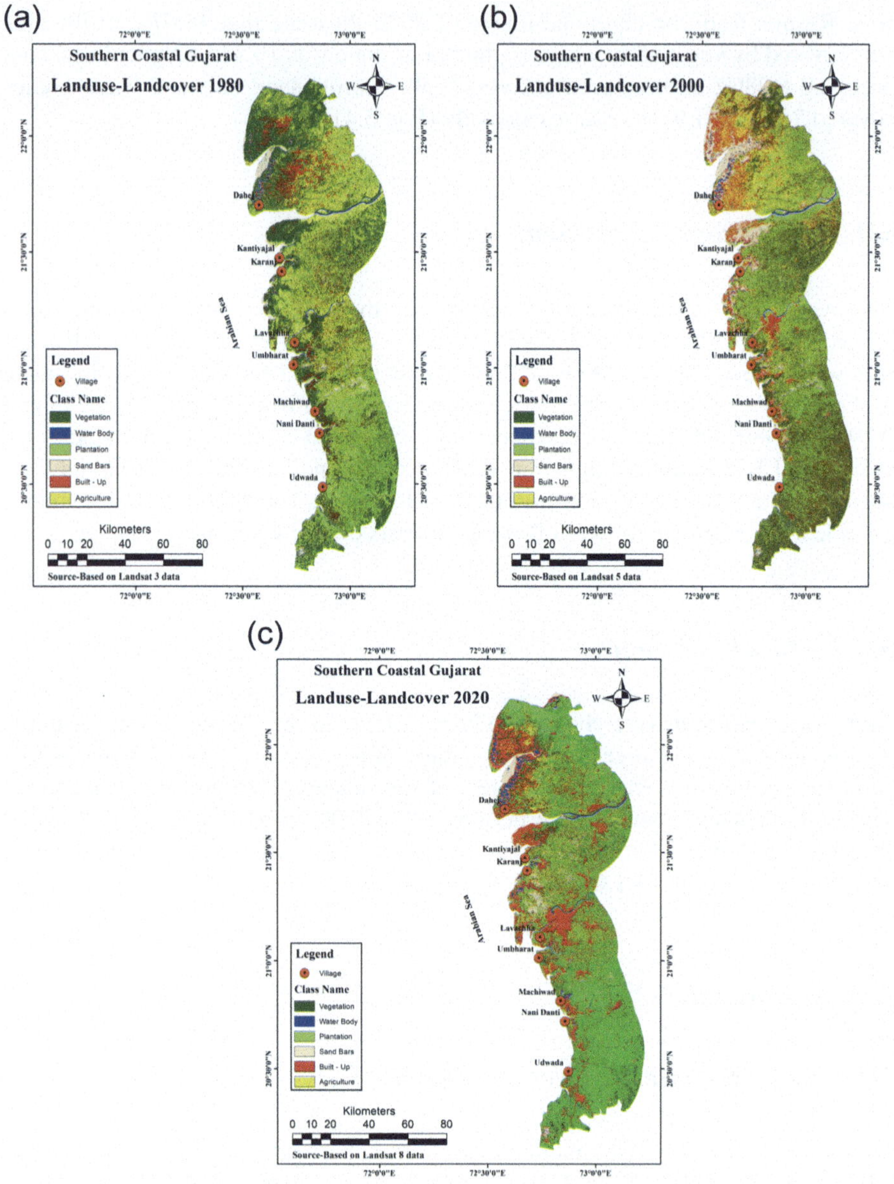

Fig. 5.3 **a**, **b** and **c** shows land use and land cover change in different categories during years 1980, 2000 and 2020 in southern coastal Gujarat

Fig. 5.4 Photographs from the field study showing changes occurring in land use and land cover

due to better access of daily human needs. The expansion is due to urban sprawl and developmental activities like building of resorts and residential complexes along the coast. About 1.65 km^2 and 10.09 km^2 area from vegetation land were changed to agriculture and plantation, respectively. The plantation along the coastal area plays

Table 5.4 Accuracy assessment error matrix of classified image 1980

Error matrix 1980

Row ID	Object ID	Class name	Vegetation	Water body	Plantation	Sand bars	Built-up	Agriculture	Classified total
1	1	Vegetation	65>	0	5	2	12	2	86
2	2	Water body	1	32	6	13	0	7	59
3	3	Plantation	0	4	48	4	1	0	57
4	4	Sand bars	0	2	2	33	2	1	40
5	5	Built-up	12	3	0	4	36	2	57
6	6	Agriculture	0	6	7	0	5	24	42
		Reference total	78	47	68	56	56	36	341

Overall accuracy = 88.90%

Table 5.5 Accuracy assessment error matrix of classified image 2000

Error matrix 2000

Row ID	Object ID	Class name	Vegetation	Water body	Plantation	Sand bars	Built-up	Agriculture	Classified total
1	1	Vegetation	72	0	5	2	2	2	83
2	2	Water body	2	55	0	3	0	5	65
3	3	Plantation	5	4	42	2	1	0	54
4	4	Sand bars	0	2	2	54	2	1	61
5	5	Built-up	0	0	0	4	47	2	53
6	6	Agriculture	0	6	3	0	5	25	39
		Reference total	79	67	52	65	57	35	355

Overall accuracy = 83.09%

Error matrix 2000

Row ID	Object ID	Class name	Vegetation	Water body	Plantation	Sand bars	Built-up	Agriculture	Classified total
1	1	Vegetation	72	0	5	2	2	2	83
2	2	Water body	2	55	0	3	0	5	65
3	3	Plantation	5	4	42	2	1	0	54
4	4	Sand bars	0	2	2	54	2	1	61
5	5	Built-up	0	0	0	4	47	2	53
6	6	Agriculture	0	6	3	0	5	25	39
		Reference total	79	67	52	65	57	35	355

Overall accuracy = 83.09%

Table 5.6 Accuracy assessment error matrix of classified image 2020

Error matrix 2020										
Row ID	Object ID	Class name	Vegetation	Water Body	Plantation	Sand bars	Built-up	Agriculture	Classified total
1	1	Vegetation	57	1	4	2	2	4	70
2	2	Water body	6	62	0	0	0	5	73
3	3	Plantation	4	1	79	0	1	0	85
4	4	Sand bars	0	1	2	48	0	5	56
5	5	Built-up	3	1	1	4	57	2	68
6	6	Agriculture	0	6	3	0	5	16	30
		Reference total	70	72	89	54	65	32	382
Overall accuracy: 83.50%									

Table 5.7 Producers and Users accuracy assessment of 1980, 2000 and 2020 classified maps

User and producer accuracy assessment						
	1980		2000		2020	
Class name	Producer accuracy	User accuracy	Producer accuracy	User accuracy	Producer accuracy	User accuracy
Vegetation	96.88%	98.04%	86.75%	91.14%	81.43%	81.43%
Water body	54.24%	68.08%	84.62%	82.09%	84.93%	86.11%
Plantation	84.21%	70.59%	77.78%	80.77%	92.94%	88.76%
Sand bars	82.5%	58.93%	88.53%	83.07%	85.71%	88.89%
Built-up	63.16%	64.29%	88.68%	82.45%	83.82%	87.69%
Agriculture	57.14%	66.67%	64.10%	71.43%	53.33%	50%
Kappa	0.761		0.795		0.799	

Table 5.8 LULC changes matrix between 1980 and 2000 in southern coastal Gujarat

1980 LULC (km^2)								
		Agriculture	Built-up	Plantation	Sand bars	Vegetation	Water body	Grand total
2000 LULC	Agriculture	14.98	5.20	4.30	0.10	11.20	0.053	35.833
	Built-up	0.7	2.30	0.79	0.006	7.89	0.10	11.786
	Plantation	1.82	4.96	5.26	0.0054	0.76	0.0009	12.80549
	Sand bars	0.26	0.1043	0.0267	4.25	0.70	0.43	5.771
	Vegetation	3.40	6.40	7.65	4.68	43.32	0.078	65.528
	Water body	0.01	0.0002	0.0016	0.56	0.026	1.24	1.82781
		21.16001	18.9645	18.0283	9.6014	63.896	1.90109	133.5513

Table 5.9 LULC changes matrix between 2000 and 2020 in southern coastal Gujarat

2000 LULC (km^2)								
		Agriculture	Built-up	Plantation	Sand bars	Vegetation	Water body	Grand total
2020 LULC	Agriculture	6.78	6.90	7.02	0.097	0.36	0.0034	21.1604
	Built-up	3.05	7.90	1.76	0.26	5.99	0.0013	18.9613
	Plantation	1.20	1.76	6.78	0.09	9.20	0.0023	19.0323
	Sand bars	0.04	0.30	0.058	5.00	0.98	1.26	7.638
	Vegetation	2.40	8.30	10.09	5.13	41.15	0.017	67.087
	Water Body	0.01	0.008	0.0016	0.47	0.026	1.42	1.91778
		13.4701	25.16008	25.7096	11.047	57.706	2.704	135.7968

Table 5.10 LULC changes matrix between 1980 and 2020 in southern coastal Gujarat

2020 LULC (km^2)								
		Agriculture	Built-up	Plantation	Sand bars	Vegetation	Water body	Grand total
1980 LUL	Agriculture	8.51	7.70	8.55	1.20	9.82	0.1076	35.887
	Built-up	2.40	2.45	2.33	0.39	4.10	0.09	11.76
	Plantation	0.89	2.45	4.20	0.67	4.58	0.01	12.80
	Sand bars	0.02	0.21	0.105	4.88	0.87	0.34	5.77
	Vegetation	1.65	9.70	10.09	3.40	40.00	1.01	65.52
	Water body	0.001	0.0005	0.0016	0.56	0.023	1.25	1.83491
		13.4701	22.5105	25.7096	11.10	59.393	2.8076	133.5725

an important role to reduce the damages and protect coastal ecosystem, human lives by acting as protective shield during extreme natural events.

5.5.2 *LULC Patterns Over the Period of 2000–2020*

About 7.70 km^2 and 8.55 km^2 area from agriculture was converted to built-up and plantation, respectively. The reason is due to the expansion of population and development activities. Increase in plantations is mainly for reason that they play a protective role to coastal ecosystem. From the built-up area, 5.99 km^2 area was converted to vegetation land. Vegetation land of 10.90 km^2 was converted to plantation.

Table 5.11 Net changes in LULC groups in each period (km^2)

Changes in LULC categories			
Classes	Net change in 1980–2000	Net change in 2000–2020	Overall changes in 1980–2020
Agriculture	14.67	7.96	–22.41
Built-up	–7.18	–6.2	10.75
Plantation	–5.22	–6.67	12.9
Sand Bar	–3.83	–3.41	5.33
Vegetation	1.63	9.38	–6.13
Water body	0.08	–0.79	0.97

5.5.3 *Overall Gain and Loss*

Land features are undergoing uninterrupted state of transformation from their present form into another form subject to local environmental suitability and native people's requirements. Table 5.10 shows the result of change detection matrix of LULC transformations, which denotes the amount of area in LULC features transferred from pre-existing state to another in the study area from 1980 to 2020. Table 5.11 indicates overall changes from 1980 to 2020. It denotes the net change in land use and land cover. One of the major land cover classes of this area is built-up and plantation area which shows an increase in the last four decades. Area covered by built-up has increased by 10.75 km^2 from 1980 to 2020. Major reason for this change was identified as expansion of population and urbanisation process. Increase in plantation is attributed to conversion of agriculture land to plantation activity. Along the coastal belt agriculture land is interchanged with plantations which plays an important role to protect coastal land from storm surge, especially during monsoon season and erosion caused by the sea. Agriculture is one of the most important land use classes in the region. The agriculture system is characterised by mixed farming, pisciculture and livestock in the study area. In the study area, agriculture has depreciated and is transformed into other forms of land use. The overall change in agricultural area was −22.41 km^2 from 35.88 km^2 in 1980 to 13.47 km^2 in 2020.

5.5.4 *A Way Forward to Coastal Sustainability*

Monitoring land use and land cover change in the coastal region is very essential for the planners, governments and scientific community. The present study conducted in the southern coastal Gujarat is step forward towards identifying the imbalance between haphazard changes in the land cover and sustainable coastal management of the region. Rapid changes in the land use and land cover in the coastal areas make it susceptible to coastal erosion and other environmental hazard. The increase in the unplanned built-up area along the coast adds to the vulnerability for the communities living near it. Vulnerability is also indicated by increase in the level of water body which is denoted by rising levels of ocean water. The decline in coastal vegetation and forest is attributable to the absence of an integrated coastal landscape management plan. Decrease in the agriculture and vegetation land cover becomes a concern as it is the foundation of the local economy. Therefore to address such kind of changing land use of coastal areas, it is necessary to come up with acceptance of sustainable development principle.

Although neither continuous nor uniform, the erosion that has been seen in this region of Gujarat is significant. The intricate interplay between river flow, waves and tides has caused significant land loss to be seen, particularly near Valsad. This erosion is caused by sediment washing away from the banks. Overall, over the research periods, erosion outweighed accretion by a wide margin. Structures along the coast were severely damaged in numerous locations due to storm surges and the waves' relentless hammering. Due to the waves' transferring sand from beaches, dunes and weakening of coastal cliffs, the shoreline also experienced severe coastal erosion. Numerous people's means of subsistence have been seriously disrupted in portions of the Navsari and Valsad beaches that are prone to erosion. Finding of the survey pointed out that numerous families were compelled to relocate in order to escape the effects of coastal erosion causing changes in the usage of land because there were no effective programmes for the sustainable development of land. Despite the fact that there is still little action being made to address environmental change affecting land use, there is enough evidence at the global and regional levels to show that changes to the coastal lands have a significant impact on many people's livelihoods and push them to leave if proactive measures are not taken.

Considering, the increasing pressures on the coastal land and unacceptable damage to the coastal zone systems, it is understandable that more environmental problems in coastal management will be attributed due to the lack of integration in coastal ecosystems for sustainability. There is a need for an integrated coastal zone management (ICZM) approach, regarding all aspects of the coastal zone, including geographical and political boundaries, in an attempt to achieve sustainability. Integrated coastal zone management provides the best use of coastal land resources in the least costly way (OECD, 1993). ICZM clarifies the roles and responsibilities of various organisations, including non-governmental organisations, governmental organisations, universities and other allied agencies for planning of coastal zones. ICZM can be understood as a tool for developing a common coastal region and policy. It can focus on resolving major disasters affecting coastal landscape such as coastal erosion. It also focuses on sustainability of the region and public participation concepts that now include improvement of coastal resilience. Therefore, it is important to breaks from previous single-minded coastal region management approaches and develops an integrated coastal region land use planning strategy.

5.6 Conclusion

The study is conducted to examine the land use change in the coastal zone of southern Gujarat and the primary factors that underlie these changes can be taken into consideration to assess the future land use configuration in the coastal zone. In the study area, it is evidently observed that the many LULC features have dynamically changed to different land uses due to natural and anthropogenic activities based on the local environmental suitability and needs of the local population. Still, much of the land cover features are transformed into built-ups without seeing their damaging impacts on the coastal systems and processes. Information about land use change is particularly important to understand the behaviour of land uses related to human activities in the study area. Moreover, information of future land use change obtained from this research can be used in supporting the arrangement of coastal zone management plan, charting future activities and guiding the steps required to achieve coastal planning aims. Appropriate coastal land use planning to attain a sustainable future is important in the coastal zone because the region still encounters several problems, including the implementation of regulations and the arrangement process of coastal management plan.

References

Anderson, J.R., Hardy, E.E, Roach, J.T, & Witmer, R.E. (1976). *A Land Use and Land Cover Classification System for Use with Remote Sensor Data. Geological Survey Professional Paper No. 964*. U.S. Government Printing Office, Washington DC

Arora, K. M., & Mathur, S. (2001). Multi-source classification using artificial neural network in a rugged terrain. *Geocarto International, 16*(3), 37–44.

Ayad, Y. M. (2004). Remote sensing and GIS in modeling visual landscape change: A case study of the northwestern arid coast of Egypt. *Landscape and Urban Planning, 73*(4), 307–325.

Baby, S. (2015). Monitorig the coastal land use land cover changes of Kuwait from spaceborne Landsat sensors. *Indian Journal of Geo-Marine Sciences, 44*(6), 1–7.

Benoit, M., & Lambin, E. F. (2000). Land-cover-change Trajectories in Southern Cameroon. *Annals of the Association of American Geographers, 90*(3), 467–494.

Brown, J. F., Loveland, T. R., Ohlen, D. O., & Zhu, Z. (1999). The global land-cover characteristics database: The user's perspective. *Photogrammatic Engineering and Remote Sensing, 65*(9), 1069–1074.

Butt, A., Shabbir, R., Ahmad, S. S., & Aziz, N. (2015). Land use change mapping and analysis using Remote Sensing and GIS: A case study of Simly watershed, Islamabad, Pakistan. *The Egyptian Journal of Remote Sensing and Space Science, 18*(2), 251–259.

Chandrasekar, N., Cherian, A., Rajamanickam, M., & Rajamanickam, G. V. (2001). Influence of Garnet sand mining on beach sediment dynamics between the Periathali and Navaladi coast, India. *Journal of Indian Assocation of Sedimentologists, 20*(2), 223–233.

Chander, G., Markham, B. L., & Helder, D. L. (2009). Summary of current radiometric calibration coefficients for Landsat MSS, TM, ETM+, and EO-1 ALI sensors. *Remote Sensing Environment Journal, 113*(5), 893–903.

Chauhan, H. B., & Nayak, S. (2005). Land use/land cover changes near Hazira region, Gujarat using remote sensing satellite data. *Journal of Indian Society of Remote Sensing, 33*(3), 413–420.

Dimyati, M., Mizuno, K., & Kitamura, T. (1996). An analysis of land use/cover change using the combination of MSS Landsat and land use map: A case study in Yogyakarta Indonesia. *International Journal of Remote Sensing, 17*(5), 931–944.

El Bastawesy, M. (2014). Hydrological Scenarios of the Renaissance Dam in Ethiopia and Its Hydro-Environmental Impact on the Nile Downstream. *Journal of Hydrologic Engineering*. https://doi.org/10.1061/(ASCE)HE.1943-5584.0001112

Fan, F. L., Wang, Y. P., & Wang, Z. S. (2008). *Temporal and spatial change detecting (1998–2003) and predicting of land use and land cover in core corridor of Pearl* River Delta (China) by using TM and ETM+ images. *Environmental Monitoring and Assessment, 137*, 127–147.

Fenglei, F., Yunpeng, W., & Zhishi, W. (2008). Temporal and spatial change detecting (1998–2003) and predicting of land use and land cover in Core corridor of Pearl River Delta (China) by using TM and ETM+ images. *Environment Monitoring and Assessment Journal, 137*, 127–147.

INCOIS. (2009). *Report on use of satellite data for detection of violation of land use along the coastal regulation zone and impact of port structures on shoreline changes*

Jaiswal, R. K., Saxena, R., & Mukherjee, S. (1999). Application of remote sensing technology for land use/land cover change analysis. *Journal of Indian Society of Remote Sensing, 27*(2), 123–128.

Kaliraj, S., Chandrasekar, N., & Peter, T. S. (2015). Mapping of coastal aquifer vulnerable zone in the south west coast of Kanyakumari, South India, using GIS-based DRASTIC model. *Environmental Monitoring and Assessment, 187*(1), 4073.

Li, Y., Zhu, X., Sun, X., & Wang, F. (2010). Landscape effects of environmental impact on bay-area wetlands under rapid urban expansion and development policy: A case study of Lianyungang China. *Landscape and Urban Planning, 94*(3), 218–227.

Luong, P. T. (1993). The detection of land use/land cover changes using remote sensing and GIS in Vietnam. *Asian-Pacific Remote Sensing Journal, 5*(2), 63–66.

Lu, D., Mausel, P., Brondizio, E., & Moran, E. (2004). Change detection techniques. *International Journal of Remote Sensing, 25*(12), 2365–2401.

Mahapatra, M., Ratheesh, R., & Rajawat, A. S. (2013). Shoreline change monitoring along the South Gujarat coast using remote sensing and GIS techniques. *International Journal of Geology Earth and Environmental Science, 3*(2), 115–120.

Meyer, W. B., & Turner, B. L. (1994). *Changes in Land Use and Land Cover: A Global Perspective* (pp. 320–342). Cambridge University Press.

Misra, R. M., Murali, P., & Vethamony. (2013). Assessment of the land use/land cover (LU/LC) and mangrove changes along the Mandovi- Zuari estuarine complex of Goa India. *Arabian Journal of Geosciences, 8*(1), 267–279.

Murali, R. M., & Kumar, P. K. (2015). Implications of sea level rise scenarios on land use/land cover classes of the coastal zones of Cochin India. *Journal of Environmental Management, 148*, 124–133.

Muttitanon, W., & Tripathi, N. K. (2005). Land use/land cover changes in the coastal zone of Ban Don Bay, Thailand using Landsat 5 TM data. *International Journal of Remote Sensing, 26*(11), 2311–2323.

Nemani, R. R., & Running, S. W. (1995). Satellite monitoring of global land cover changes and their impact on climate change. *Climate Change, 31*, 395–413.

Nicholls, R. J., & Small, C. (2002). Improved estimates of coastal population and exposure to hazards. *American Geophysics Union, 83*(28), 301–305.

OECD. (1993). *Coastal Zone Management—Integrated Policies*

Raju, N. S. N., & Kumar, V. S. (2006). Coastal processes along the Indian coastline. *Current Science, 91*(4), 530–536.

Riebsame, W. E., Meyer, W. B., & Turner, B. L. (1994). Modeling land-use and cover as part of global environmental change. *Climate Change, 28*, 45–64.

Sabins, F. F. (1987). *Remote Sensing: Principles and Interpretation* (pp. 312–320). W.H. Freeman and Company.

Saha, A. K., Arora, M. K., Csaplovics, E., & Gupta, R. P. (2005). Land cover classification using IRS LISS III image and DEM in a rugged terrain: A case study in Himalayas. *Geocarto International, 20*(2), 33–40.

Shalaby, A., & Tateishi, R. (2007). Remote sensing and GIS for mapping and monitoring land cover and landuse changes in the Northwestern coastal zone of Egypt. *Applied Geography, 27*(1), 28–41.

Weismiller, R. A., & Momin, S. J. (1977). Change detection in coastal zone environments. *Photogrammatic Engineering and Remote Sensing, 43*(12), 1533–1539.

Wu, S. Y., Yarnal, B., & Fisher, A. (2002). Vulnerability of coastal communities to sea-level rise: A case study of Cape May County, New Jersey, USA. *Climate Research, 22*, 255–270.

Yagoub, M. M., & Kolan, G. R. (2006). Monitoring coastal zone land use and land cover changes of Abu Dhabi using remote sensing. *Journal of Indian Society of Remote Sensing, 34*(1), 57–68.

Yuan, F., Sawaya, K. E., Loeffelholz, B., & Bauer, M. E. (2005). Land cover classification and change analysis of the Twin Cities (Minnesota) Metropolitan Area by multi temporal Landsat remote sensing. *Remote Sensing Environment Journal, 98*, 317–328.

Zoran, M. E. (2006). The use of multi-temporal and multispectral satellite data for change detection analysis of Romanian Black Sea Coastal Zone. *Journal of Optoelectronics and Advanced Materials, 8*, 252–256.

Zhang, R., & Zhu, D. (2011). Study of land cover classification based on knowledge rules using high-resolution remote sensing image. *Expert Systems with Applications, 38*(4), 3647–3652.

Chapter 6
Interplay Between Tidal Range and Dam-Induced Flooding in Sustaining the Islands and Livelihoods in Narmada Estuarine Zone, India

Saransh Kumar

Abstract Development of channel cut-off estuarine islands is a complex phenomenon influenced by various factors. This study focuses on understanding the interplay between channel discharge and tidal surge in causing morphological variations in lower fractal peninsular rivers. Using the case study of Taria Island in the Narmada Estuary, we investigate the temporal development of the island and river bank line changes. This study underscores the importance of considering multiple factors, including tidal energy and sediment dynamics, in understanding the formation and evolution of estuarine islands. The integration of remote sensing techniques and field observations allows for a comprehensive analysis of the complex processes driving morphological changes in estuarine environments. Multispectral and RADAR-based remote sensing data at 30 m resolution are utilized to assess the role of tidal influx in facilitating the convergence of suspended sediment load. The Normalized Difference Water Index (NDWI) is employed to illustrate the seasonal mechanism of sediment stabilization in the channels. Our findings reveal an inverse relationship between stream discharge during non-monsoon months and sediment accumulation attenuated by tidal pivoting. Through the interpretation of remotely sensed data, we conclude that the large amplitude of the tide in the Gulf of Khambhat region provides the necessary energy for sediment accretion in the channels south of the island, resulting in a raised riverbed on one side and contributing to the enhancement of the island area. These findings contribute to a better understanding of the interactions between hydrological and geomorphological processes in estuaries, highlighting the significance of tidal energy in shaping estuarine landforms and promoting sediment deposition.

Keywords Geo-hydrology · Sediment flux · RADAR · Estuarine Island · Tidal influx · Pivotal circulation

S. Kumar (✉)
National Institute of Advanced Studies, IISC Campus, Bengaluru 560012, India
e-mail: saransh@nias.res.in

© The Author(s), under exclusive license to Springer Nature Switzerland AG 2024

B. V. Raj Singh and A. K. Batar (eds.), *Sustainable Local Development for Environmental and Social Sustainability*, Human-Environment Interactions 11,
https://doi.org/10.1007/978-3-031-67303-0_6

6.1 Introduction

Floodwaters are often perceived as disastrous due to their negative impacts on life and property in a region. However, from a geophysical perspective, floods serve as a means for rivers to alleviate short-term distress. Graf (1987), (1997) describes rivers as manifestations of the interactions among three landscapes: a landscape of energy, a landscape of resistance, and a landscape of geomorphic work. These interactions, represented by spatial distributions of relevant physical measures, give rise to dynamic water and sediment flow processes that shape distinct morphological features. This dynamism can range from the formation of puddles in riverbeds, causing micro-scale vorticities, to large-scale embankments or the development of sandbars (Allen, 1997).

Slater (2019) demonstrated how climate variability influences river channel transport in alluvial rivers, where channel densities may increase during extensive water flow and narrow down during periods of below-normal flow. Macro-scale changes, such as meandering, may decrease or increase depending on flow velocity and density as empirically demonstrated in channel migration studies (Njenga, 2012). Due to the continuous state of dynamism, measuring changes in rivers using on-field methods has limitations in time series analysis. Current methods are insufficient to provide a theoretical explanation, especially regarding flow dynamics in non-ideal systems (Jasper Leuven, 2018).

In this study, we aim to explain the morphological changes that have occurred in the Narmada estuary over the past 30 years using remote sensing data and change detection through time series analysis. It has been established that while there is a tendency for curvature change relative to flow velocity, the sinuosity of the pre-flood (or normal channel morphogenesis) can affect the river water vectors (Graf, 2006). At the point of discharge into the sea, estuary, or delta, the tidal forces significantly impact these flow vectors, consequently affecting sediment discharge (Dronkers, 1986). The influence of tides can be extensive in deltaic rivers, whereas in estuaries with small inlets, this opposing force tends to be more focused and variable. Erosional and depositional forces are both active with large variations, as demonstrated by Nayak and Sahai (2007) in their study on the effects of tides on suspended sediment concentration, leading to a diverse range of morphological features.

There is limited literature currently existing in the field domain discussing the formation of channel-cut islands and the mechanisms involved in their repair over time and limited data on the role of waves and tides in sediment transport (Davies, 1964). These unique fluvial mechanisms raise questions about the underlying phenomena which sustain the morphological landforms developed in the river bed. While the mechanism of tides in the Narmada estuary region has been extensively researched (Harish Gupta, 2005; Kumar & Kumar, 2010; Srinivas, 2005), rarely have these phenomena been utilized to explain sediment transport and its extensive reach upstream in the river. The highly variable channel flow due to geo-hydrological barriers in the senile stage of the river makes them particularly interesting for study. The lower course of the Narmada River exhibits such adjustments, giving rise to peculiar landforms. Thus, this study primarily aims to utilize remotely sensed data to explain and demonstrate:

(a) The stages of development and extent of modifications of Taria Island in the Narmada estuary.
(b) The mechanism that explains the role of tidal energy and channel shifting processes in simultaneous incision and sediment accumulation along the riverbank.

The paper first provides the study area for the region where competing processes of erosion and deposition take place followed by approach followed to capture the changes observed in the river bed over time remotely. We provide results in the following sections on accounts of the change in shape of the island and quantify the accretion and deposition in terms of changes in the areal extent of the island. We further provide an analysis of this change by explaining the probable reasons for the physical processes and the sustainability of people dependent on it.

6.2 Study Area

The Taria Island, located in the Narmada Estuary within the Bharuch District of Gujarat, exhibits a unique process of erosion, accretion, and sedimentation that occurs simultaneously and competitively. The island's coordinates are approximately 21°67′ N and 72°87′ E. The island marks the end transect of the river for administrative delineation if not physical. The study area of interest (AOI) surrounding Taria Island is bounded by 21°38′30″ to 21°41′30″ N and 72°49′30″ to 72°55′30″ E, focusing on analyzing the changes in the island's extent. The turbid state of water is found several kilometers inland from the island in the river principally due to the large amplitude and range of the high tide in the Gulf of Khambhat.

The Narmada River, otherwise known as Reva, Samodbhava and Mekalasuta (Gaur, 2016) is one of the largest west-flowing rivers, with its regime determined by the fault geomorphology of the Peninsula leading toward the Arabian Sea (Chisholm, 1911). The Sardar Sarovar Dam has been constructed approximately 120 km upstream of the study site. This dam also marks the transition of the river to a senile stage, characterized by a dramatic increase in sinuosity index. The width of the river channel at the dam site reaches 488 m during high floods and 45.70 m during the summer season. The maximum recorded flood on September 7th, 1994, was 70,847 cumecs, while the minimum recorded flow in summer was 8.5 cumecs. The mean annual rainfall in the basin is 112 cm, with the majority occurring from August to October (IMD citation needed). The annual runoff through the dam transect site, with 75 percent dependability, is 27.22 million acre-feet (MAF), as computed by the World Bank (SSNNL, updated July 2022), which estimated the yield as 28.57 MAF (Fig. 6.1).

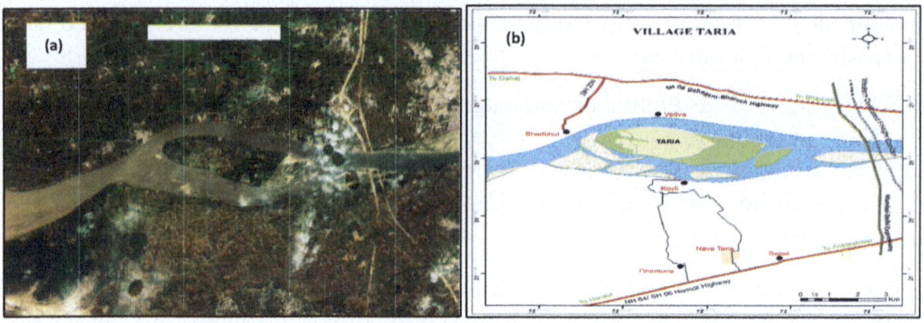

Fig. 6.1 (from left) Thematic map of the island Taria; RGB true color imagery of the estuarine island vis a vis relief association. *Source* **a** Bhuvan ISRO; **b** Sentinel 2 (Google Earth Engine)

6.3 Methodology

The study employed a time series analysis using remote sensing techniques and multi-spectral data captured in the Near-Infrared (NIR) and Shortwave Infrared (SWIR) bands. The study focused on analyzing the morphology of the Taria Island and the sand and mud flat deposits in the surrounding channel, with data collected during four phases at ten-year intervals (1991, 2001, 2011, and 2021). Bands 4 and 5 (Landsat 5 and 7) for 1991 and 2001 and Bands 5 and 6 (Landsat 8) for years 2011 and 2021 were used. [Refer to Table 6.1 for sensor cross-calibration]. To minimize the influence of tide phase and seasonality on the analysis, selective images representing the first low tide phase of the river and captured during the months of March and April, prior to the monsoon season, were chosen. The time of image capture was set between 10.20 and 10.25 h, and the closest available images to this time interval were selected. Fine siltation and coarse sediment flats were further segregated based on the differences in the wavelength spectral profile using density slicing. Following this, temporal boundaries were traced using ArcGIS (ArcMap 10.5) platform to quantify the extent of area change and shoreline recession of the river.

To show the seasonal deposition of the sediments and differentiating mobile sediments from stationary sand flats, a Normalized Differentiated Water Index (NDWI) was computed using Map Algebra tool of Spatial Analyst toolbox given by Rousse et al. (1973) and applied in part by Mohammad Hadian 2021):

$$\text{NDWI} = \frac{\text{Band 2(Green)} - \text{Band 4(NIR)}}{\text{Band 2(Green)} + \text{Band 4(NIR)}}$$

Seasonal (Level 2 pre-calibrated) images for phases (pre-monsoon, monsoon, and post-monsoon) were obtained from ResourceSat 1 LISS III sensor at 23.5 m resolution, thus the syntax used in the processing NDWI varies. All the images used were chosen at similar resolution levels and band selection for specific type of analysis. The following table explains the different bands of the sensors used in the analysis.

Table 6.1 Data used from different sensors in the various stages of analysis of the images

	Mission	Source	Sensor	Year	Bands & spatial resolution	Bands selection
1	Landsat 5	Earth explorer USGS	TM	1990–91	Band 1–5, 7 30 m Band 6 120 m	B4 NIR (0.76–0.90 μm) B5 SWIR1 (1.55–1.75 μm)
2	Landsat 7	Earth explorer USGS	ETM+	2000–01	Band 1–5, 7 30 m Band 6 60 m Band 8 15 m	B4 NIR B5 SWIR1
3	Landsat 8	Earth explorer USGS	OLI	2010–11, 2020–21	Band 1–7, 9 30 m Band 8 15 m Band 10–11 100 m	B5 NIR B6 SWIR1
4	Resourcesat 1	Bhuvan ISRO	LISS III	2017 (Seasonal)	Band 2–5 23.5	B3 Green B4 NIR
5	Cartosat 1	Bhuvan ISRO	PAN Stereo	2008	PAN DEM 2.5 m DSM 30 m	

A land cover change map was generated for the AOI to assess changes in land cover and land use patterns since 1991. Supervised classification using the maximum likelihood method was employed to differentiate pixel tone changes. Radiometric correction was applied to the images for accurate change detection. The results were validated using secondary ISODATA clustering.

To understand tidal current and sediment transport mechanisms, gridded data from Oceansat 2 OSCAT and SCATSAT sensors, as well as tide wavelength data from the National Hydrographic Office, were analyzed. Monthly tidal wave plots were generated to assess wavelength variation during consecutive high and low tides based on the lunar cycle.

A digital surface model with 30 m resolution was created using Cartosat 1 PAN 2.5 m stereo data to study river bed topography, including roughness and channel disruption. The elevations were cross-validated with SRTM data at the corresponding resolution.

6.4 Results and Discussion

6.4.1 Temporal Changes in the Study Area

Near-infrared (NIR) and short-wave infrared (SWIR) images used to distinguish the sediment accumulation over the island and surrounding morphological features like mudflats over a period of 30 years are shown below [Fig. 6.2 composite]. The spectral bands between 0.9 and 2.7 μm were particularly effective in distinguishing the land–water boundary and differentiating the sediment load (stationary) from the dissolved or suspended sediments (mobile load) in the water.

Through the analysis of time series data (1990–2020), it is seen that the mud flat accumulation has been accelerating on the southern bank of the island. From recent bathymetric data analyzed in related studies (Harish Gupta, 2005), the sediment supply in and around the island is attributed to the above-average flood discharge of the river during the monsoon season. The narrow northern channel experiences swift water currents (>5 m/s) (Sinha, 2010), suggesting the dominant river channel lies toward the north of the island, consistent with sediment transportation in open channels (Graf, 1984). This corroborates the tidal mixing study in Dahej Creek at the mouth of the river in the study by Narayana Swamy and Sarma 1982). The images illustrate the development of pea-shaped mud flats with pointed tips in the southern section, while no sediment accumulation features are observed in the northern channel.

The wind pattern in the Gulf and the estuary, as depicted in wind vectors, demonstrates a general flow of winds (Fig. 6.2b). The wind pattern in this area is predominantly seasonal with rare cyclonic disturbances. Predominant wind directions in this area are west–south-westerly and north–north-easterly during the months of June to September and December to March, respectively. Higher wind speeds occur during June to September with winds of up to 74 km from the west and southwest (Kunte, 2008). Tidal currents get additional push to accrete the suspended sediments upstream.

Tidal flooding exhibits littoral surficial currents that follow the southern bank line of the river, carrying finer to medium-grain sediments based on the tidal amplitude during high tide phases. The tidal range and semidiurnal nature of the tide facilitate the transportation of finer sediments along the southern channel, which was previously cut by the river during flooding events. The dominant role of tidal floods along the southern bank is evident from the formation of semi-stationary sandbars and mud flats (See Fig. 6.2 left images). These sandbars exhibit pointed tips, indicating a bidirectional flow of the current.

The sedimentation along the southern bank follows the pattern of how Alia bet is getting joined with the landmass as studied by (Subash Bhandari, 2018) and may offer a similar prognosis in the pattern if the rate of positive feedback continues to reinforce the sedimentation. If the rate of sedimentation increases over and above the current rate of deposition, it may provide scope for tidal waters led accretion to augment further.

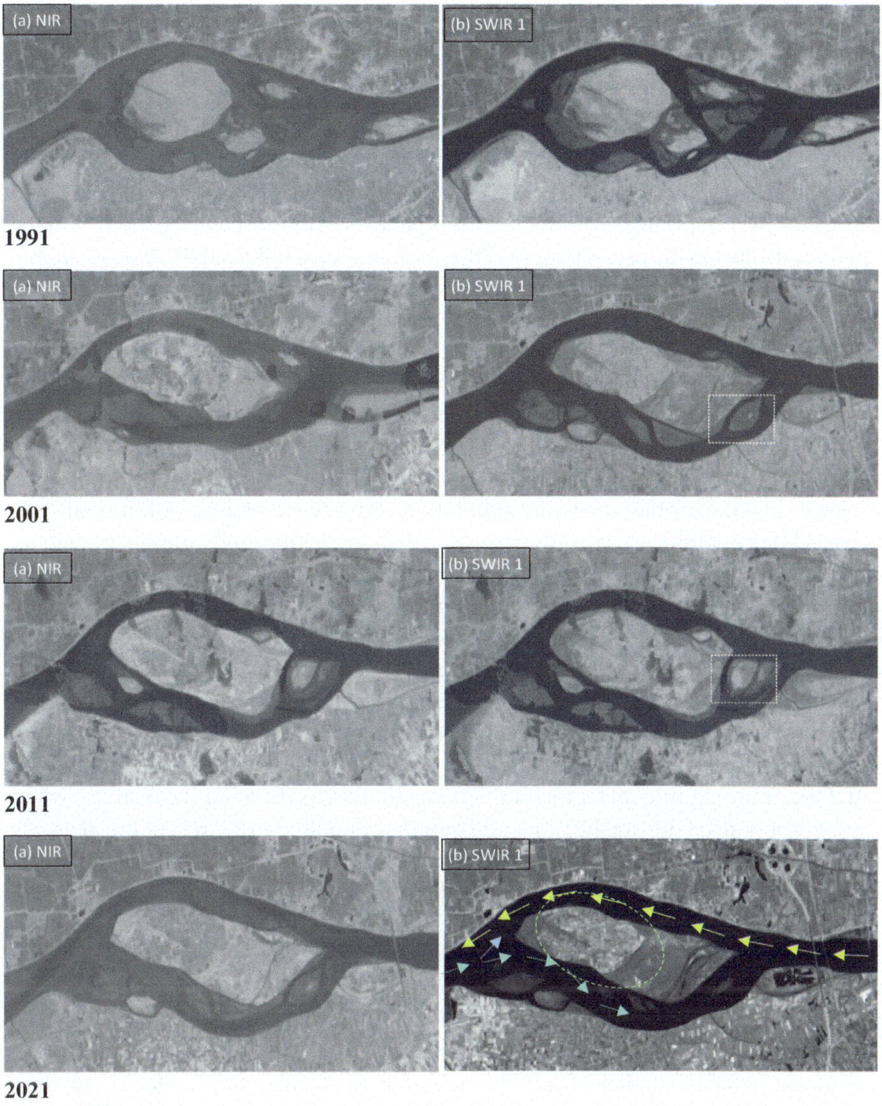

Fig. 6.2 (from the left top)—Island morphogenesis from years 1991 to 2021 showing variation in river bank, mudflats, and the dominant flow in river Narmada in the study area. Left column of images is near-infrared band and the right column represents short-wave infrared band. Data Source—Landsat Level 1 Collection, Explorer USGS Repository. Annotation in the 2021 Image B shows the general movement of water in the estuary. Yellow arrows represent the dominant channel of the river while the blue arrows represent the dominant path of water during a typical tidal high

6.4.2 Tidal Amplitude and Semidiurnal Nature of Tides

However, this short interval of flow reversal during high tide is insufficient to generate significant sediment transport, resulting in a turbid state with a hypopycnal load in the estuary. The southern channel predominantly receives tidal transport through littoral drift, while the northern channel, with greater runoff energy, serves as the dominant channel for the river. The flow velocity discrepancies in these channels, along with the high tide currents in the estuary support this explanation (Sinha, 2010). Consequently, a tidal gyre is formed during the non-monsoon season, wherein the water flow from the northern channel merges with the gulf, creating a pivotal circulating path during high tide. During the monsoon period, when the river runoff exceeds its average volume, this tidal gyre either weakens or breaks down to an extent where significant sedimentation related to tidal transport is not prominently visible.

Due to the absence of a data collection center in the estuary and the closest tidal station being located at Suvali (Hazira Port), 30 km to the south, INCOIS publishes predicted tidal ranges based on tidal simulation and wave modeling. Additionally, the National Hydrographic Office publishes tidal data for fishing advisories in the estuary, which are used for reference in the explanations (Fig. 6.3a, b). As depicted in the graph, apart from the daily extremes, the tidal wavelength in a typical month strongly aligns with the lunar cycle. Higher amplitudes provide sufficient potential energy for currents to transport fine to medium sediments upstream in the estuary, serving as the primary driving force behind the propagation of accretionary sedimentation.

In the specific case of the Narmada estuary, it is crucial to establish a regional circulation framework to understand the movement of water and sediments into the Gulf. One of the significant factors influencing this site is the wind shear stress, which is more pronounced during the monsoon months compared to the rest of the year. The primary lunar component of the tide contributes to the largest anomaly in amplitude in the Indian subcontinent and is likely the defining variable that explains the resulting tidal height in the region (Srinivas, 2005). The tides in the Gulf of Khambhat remain the most substantial along the Indian coast, with an average spring tidal range of approximately 9 m above mean sea level. Considering the bed height in the study area, ranging between 3 and 5 m in AOI, the resulting wave amplitude is sufficient to generate strong currents (Kumar & Kumar, 2010).

Srinivas (2005) established a relationship between tidal amplitude, along-shore water currents, and their temporal variability during different seasons. The larger gravitational pull and release during tides provide the water with sufficient energy to travel over a greater distance while carrying a significant sediment load (refer to Fig. 6.4I). When combined with seasonal wind shear stress (see Fig. 6.4II) and the shape of the river mouth, additional constraints are imposed on wave propagation, directing the energy upstream (Unnikrishnan et al., 1999).

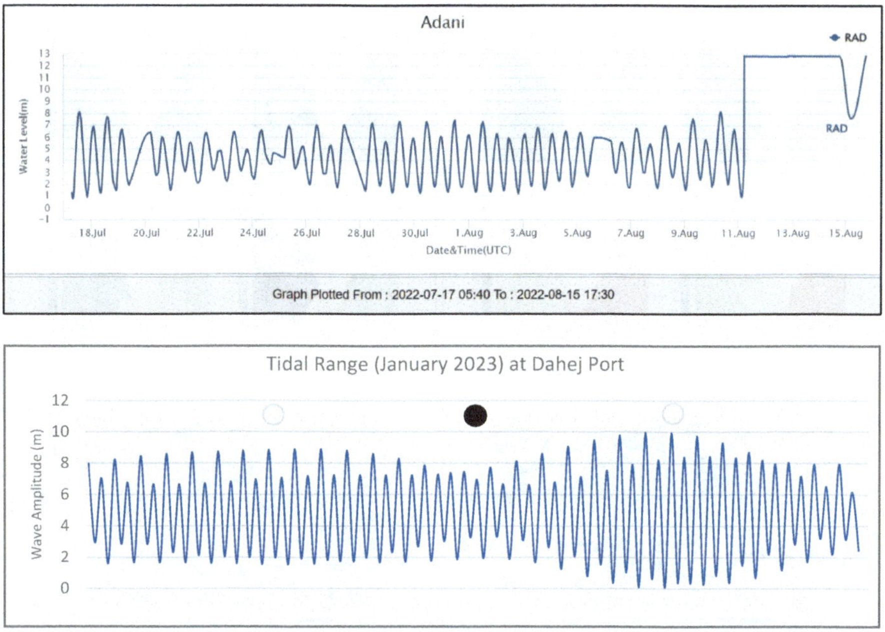

Fig. 6.3 (top) Recorded sea level heights at Adani Port, Hazira 50 km at nearest intersect and (below) Simulated tidal range at Dahej Port (5 km west-northwest to the study area). *Data Source* (Above)—INCOIS; (Below)—National Hydrographic Office, Dehradun

6.4.3 Channel Flow, Bed Topography, Channel Load Shifting Process

The flow of water in the Narmada River has been significantly influenced by geo-hydrological barriers, leading to frequent shifts in its path. There are two main reasons contributing to these shifts. Firstly, the drastic changes in relief play a significant role. The location of the Sardar Sarovar dam upstream marks the end of the rift zone control at an altitude of 110 m above mean sea level (MSL). Beyond this point, the flow of water and the elevation decrease toward near mean sea level (between 30 and 10 m above MSL) (refer to Fig. 6.5). At the study site, the elevation ranges from 3 to 5 m above MSL. Secondly, the reduced mean water flow and infrequent release of large volumes of water downstream of the dam during the monsoon season causes erosion, but the sediments do not necessarily get transported to the sea (Gupta, 2014). Instead, they accumulate as multiple levees in the river channel. This accumulation of sediments contributes to enhanced meandering of the channels and occasional processes of incision and channel capturing.

Establishing the bathymetry of the estuary is challenging due to the limitations of available remote sensing sensors. Since LIDAR applications are either unavailable or restricted in the study area, the analysis draws inference from studies conducted

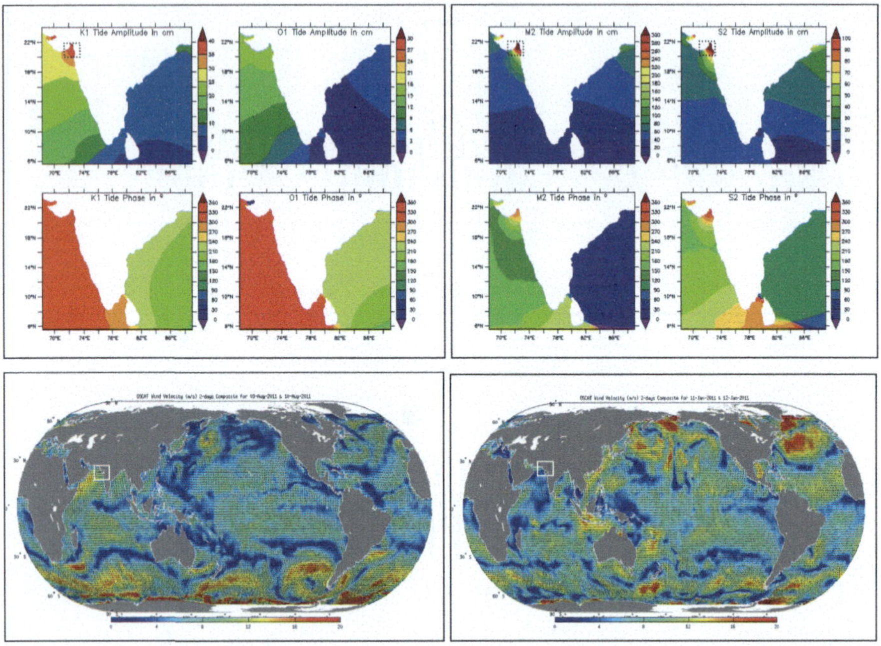

Fig. 6.4 (I) from left—K1 and O1 tide components in the Khambhat region showing significant concentration of lunar principal component near estuarine zone. M2 and S2 components of tide were measured from SCATSAT sensor (extracted from NICES project on Bhuvan platform of NRSC-ISRO. (II) Oceansat OSCAT derived wind vectors at 25*25 km resolution showing enhanced wind stress in monsoon months (left) in the Khambhat region near Gujarat, India. *Data Source* Bhuvan ISRO

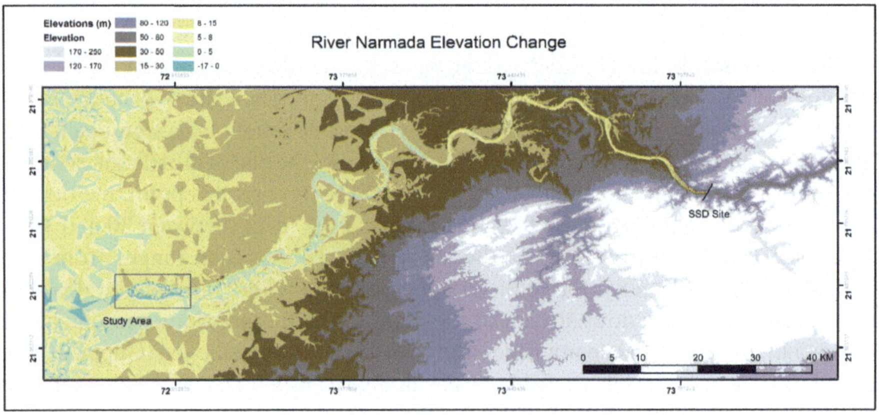

Fig. 6.5 DEM showing the relief change from the site of dam and the study area

by Sinha (2010), which utilized hydrographic surveys, wave modeling, and sediment transport dynamics studies in the region. To enhance reliability, Cartosat 1 orthorectified data was used to develop a profile of the bed topography in the study area. The riverbed is deeper near the northern channel bank compared to the southern channel, which is more prone to inundation but remains submerged.

Mudflats make the spearhead and tail of the island and most of the river flow is a function of the depth and irregularities in the bed [blue pixels near island inside the river channel in Fig. 6.6). Numerous channels are present in the southern section due to high siltation accreted by the tidal flood and ebb process. The dominant larger channel is present north of the island. In addition to the bed irregularities, the dominant flow through the northern channel is contributed by both deep pools and the meandering of the channel just before the island providing necessary angle to focus the flow of water. The current topography of the river section shows bed roughness in the northern section with narrow flow deepened toward the northern bank of the river. Here, the flow provides a case for Bernoulli's principle which states that the speed of the fluid increases dramatically with simultaneous decrease in fluid pressure.

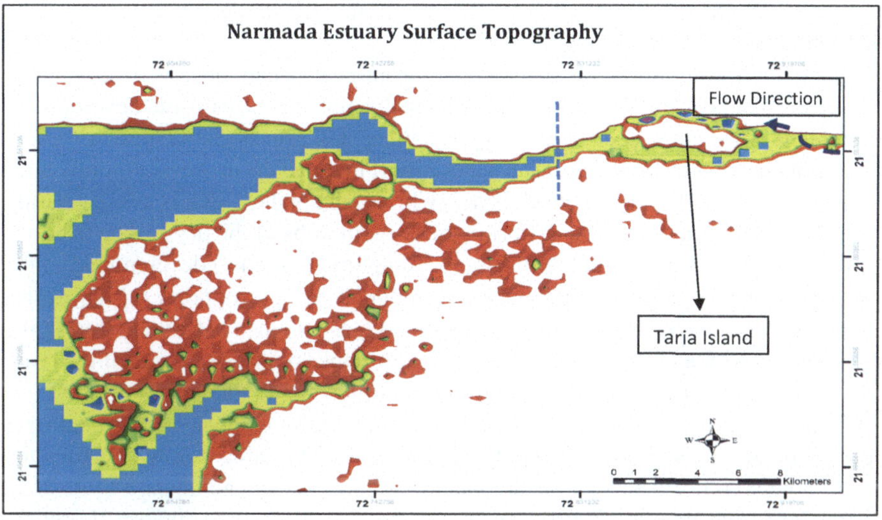

Fig. 6.6 River bed topography of the river (DSM extracted at 30 m resolution using Cartosat PAN stereo). {Pink – depth = > -3 m, Magenta = between −1 and −2 m, Blue = Between MSL upto −1 m, Yellow 0–2 m above MSL, Red 2–5 m above MSL}. Data Source-Bhuvan ISRO. (Blue pixels after dotted line do not represent any information because of image processing limitation)

6.4.4 Island Morphology and Sedimentation

River sedimentation increased drastically in the downstream section of the dam after the Sardar Sarovar Dam (SSD) was built. The river and its tributaries in their entire course have five major geo-hydrological barriers which impound water and store. SSD marks the last functional barrier at present and is not needed to release water for its massive storage capacity, unlike in certain cases during excess monsoon. Thus, the water flow and the sediment transport remain largely variable. Almost 85 to 98% of annual sediment load in the river is transported during the monsoon season from June to November (Harish Gupta, 2005). This essentially reduces the river's capacity to flush out water over time, thus factoring in a part of sediment growth of the island. Pea-shaped sand bars and mud flats around the island are exposed in low tide conditions. The whole southern channel runs over a raised riverbed due to accretionary tidal flooding twice a day (channel depth difference of upto a meter). A narrow channel coupled with a bed roughness factor increases the erosive state toward the coast. It is to be noted that this process of erosion, accretion, and deposition happens simultaneously at different levels, but on a larger scale. The northern channel facilitates dominant erosion transport and the southern channel marks the area of deposition. While the accretion happens simultaneously along both coasts of the island in between.

Channel differentiation becomes more prominent when observed seasonally. Stabilization of post-monsoon water flow happens in winter season and development of mud flats is visible as seen below (Fig. 6.7 Image C). This does not however imply that the channel has a stable flow. The water runoff in the rivers decreases rapidly after the rains enhancing the range of tidal energy in the channels. There is usually hardly any or no runoff into the estuaries during the eight months of the dry season (Shetye, 2011). This affects the channel flow largely. Temporal (seasonal) development of the flow flux shows an augmented state where reduced energy supports further accretion along the southern channel. The sedimented bed becomes prominent during the pre-monsoon with a narrow channel incised within the southern channel of the river. The degree of curvature suggests a high sediment load and reduced water flow.

Competing processes have resulted in changes in the extent of the island over a period of 30 years, starting from 1990 (Fig. 6.8). The table below provides the quantifiable significant growth of the island and the percentage change in its area. It is evident from the data that the river continues to erode the southern shoreline, while sediment deposition on the island leads to a significant elongation of sand flats, contributing to the overall increase in the surface area of the island. A similar geospatial approach (Chirala & Pedada, 2021) regarding Meghadrigedda reservoir near Visakhapatnam, India has been conducted which shows time series changes in locations of erosions in the catchment area.

Between 1991 and 2011, the size of the island has increased by 69.14%along with an increase in the density of mudflats and sand bars in the surrounding southern channel. Along with channel deposition, the southern shoreline has also witnessed recession. The change is approximately 128 m every decade since 1991. While the

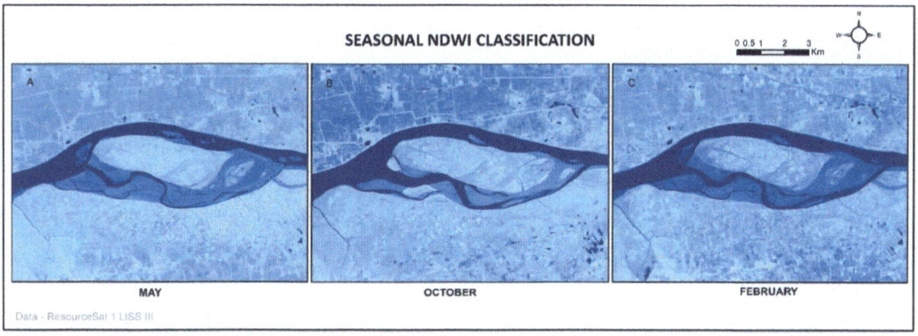

Fig. 6.7 Normalized Differentiated Water Index showing seasonal stabilization of sediments brought in by tidal flood south of the island. In post-monsoon (October) phase the sediment is relatively mobile than in dry months of February and May

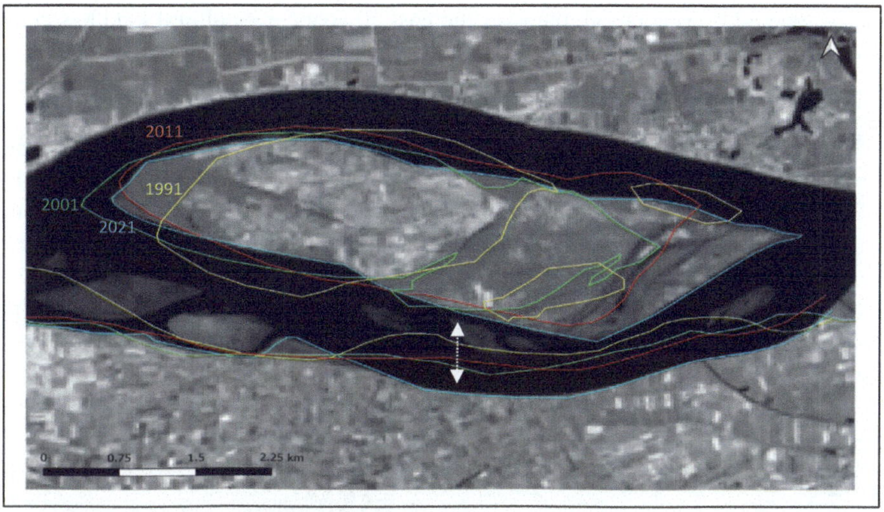

Fig. 6.8 Temporal extent of the island and the change in the shoreline shown in respective colors overlayed on a short-wave infrared imagery (Band 6, Landsat 8 OLI) (Year—2023). (The extent of the island was delimited by developing a differentiated pixel profile)

island has shown signs of stabilization owing to decreasing rate of increase over the decades, the shoreline continues to show dynamism (Table 6.2).

Table 6.2 Temporal (decadal change) extent of sedimentation and erosion along the island. (Figures in parenthesis are in percentage)

Year	Island extent (km^2)	Increase in area km^2 (%)	Shoreline recession (in m)
1991	4.980	Baseline	Baseline
2001	7.223	2.243 (45.04)	162.6
2011	8.382	1.159 (16.05)	78.8
2021	8.423	0.041 (0.49)	142.5

6.4.5 *Sustainability of Livelihoods and Change in Land Use Pattern*

The significant morphological changes in the region have also led to alterations in the land cover pattern, as depicted in (Fig. 6.9). Previously, the island was predominantly covered by vegetation, but over the past 30 years, agricultural land has been developed. Additionally, scattered settlement structures have emerged, indicating a sparse population on the island. Infrastructure growth is also noticeable, represented by the red-colored areas in Fig. 6.8.

The relatively stable water flow in the river has facilitated seasonal cultivation on the island. However, the cropping pattern remains limited to single crops, primarily belonging to the legume and pulse family. Due to the unique position of the island, irrigation for these agricultural fields relies on the monsoonal flooding of the river, which inundates the dry fields with fresh water. It is worth noting that the water quality is more saline toward the seaside compared to the east side, owing to the east–west flow of the river. This observation from the field reinforces the significant role of tidal wash originating from the west and flowing upstream into the river.

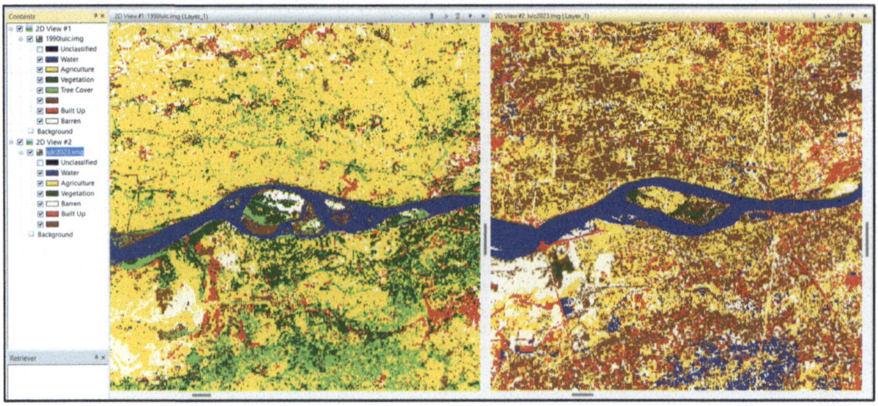

Fig. 6.9 Change in land cover pattern in between years 1990 (left) and 2023 (right). Data Source–Landsat 5 TM (for 1990)/Landsat 9 OLI (for 2023). Processing-Supervised Classification (ERDAS Imagine)

This flood channel cut-off island at present hosts a tribal community which earlier was devoid of land ownership in the village. As the forward communities migrated away from the river following risks due to flooding in the past, this land was left available when the flow of the river stabilized over the period. This land was occupied by these communities which has till now contributed to raised livelihood and quality of life for them, despite barriers to access the island. The island has converted from marshy, scrub lands to cultivable patches occupied by scattered households belonging to this community. Overall, we see a 67% rise in farmland conversion over 30 years.

6.5 Conclusion

This study sheds light on the complex interplay of various forces in shaping the Taria Island and highlights the role of sediment deposition from tidal wash and river runoff. The island's shape and size have predominantly been sustained by the accretion of sediments, leading to its aerial growth. Although the rate of sediment accretion has slowed down over the years, the southern channel of the island continues to exhibit dynamism in terms of fine sedimentation. The mud flats become more pronounced during non-monsoon months when the rate of sedimentation increases.

The tidal-influenced water current acts as a secondary mechanism for sediment transport in the estuary, supporting the deposition of sediments along both shores of the island (Singh, 2021). The differential capacity of sedimentation is evident in the southern section, which experiences bidirectional water flow during different phases of the day, as inferred from the tips of the mud flats in the southern channel. Additionally, the connection of Alia bet (formerly an island) with the mainland indicates an enhanced rate of sediment deposition over decades.

This study provides a foundation for further exploration of cases where geo-hydrological barriers trigger morphodynamic changes in estuaries. It emphasizes the role of tidal energy in repairing landmass through sedimentation and establishing a dynamic equilibrium within the system. Overall, it contributes to a better understanding of the sedimentary processes and their impact on the estuarine ecosystem and their differential impact on the livelihoods of people.

References

Allen, P. A. (1997). *Earth surface processes*. Blackwell Science.

Chirala, U., & Pedada, B. (2021). Hydrogeomorphology, NDWI and NDVI of the meghadrigedda sub-watersheds for optimal utilisation of resources, Visakhapatnam District, Andhra Pradesh—India using Landsat Data 2000 and Sentinel Data 2020. *International Journal of Geosciences, 12*(06). https://doi.org/10.4236/ijg.2021.126033

Chisholm, H. (1911). Nerbudda. In *Encyclopaedia Britannica Vol. 11*. C. U. Press, Ed. Retrieved from Encyclopaedia Britannica.

Davies, J. L. (1964). A morphogenic approach to world shorelines. *Geomorphology*, 27–42.

Dronkers, J. (1986). Tide-Induced residual transport of fine sediment. In J. van de Kreeke (Ed.), *Physics of shallow estuaries and bays* (pp. 228–244). Springer-Verlag.
Gaur, A. S. (2016). A maritime archaeological exploration in Narmada Estuary, West Coast. *Marine Environment*, 82–87.
Graf, W. H. (1984). *Hydraulics of sediment transport* (4 ed.). Water Resources Pubns.
Graf, W. L. (1987). *Fluvial processes in dryland rivers*. Springer.
Graf, W. L. (1997). Natural and anthropogenic influences in fluvial geomorphology. *Journal of Hydrology*, 165–167.
Graf, W. L. (2006). Downstream hydrologic and geomorphic effects of large dams on American rivers. *Geomorphology*, 336–360.
Gupta, M. (2014). Monitoring shoreline changes in the Gulf of Khambhat. India during 1966–2004 using RESOURCESAT—1 LISS III. *Open Journal of Remote Sensing and Positioning, 1*, 27–37. https://doi.org/10.15764/RSP.2014.01003
Harish Gupta, G. J. (2005). Temporal and spatial variations in water flow and sediment load in Narmada River Basin, India: natural and man-made factors. *Environmental Geology*, 579–589.
Jasper, R. F. W., & Leuven, S. L. (2018). Empirical assessment tool for bathymetry, flow velocity and salinity in estuaries based on tidal amplitude and remotely- sensed imagery. *mdpi*, 243–274.
Jasper Leuven, S. V. (2018). Empirical assessment tool for bathymetry, flow velocity and salinity in estuaries based on tidal amplitude and remotely-sensed imagery. *Remote Sensing, 10*(1915), 243–274. https://doi.org/10.3390/rs10121915
Kumar, V. S., & Kumar, K. A. (2010). Waves and currents in tide dominated location off Dahej, Gulf of Khambhat, India. *Marine Geodesy, 33*, 218–231.
Kunte, P. D. (2008). Sediment concentration and bed form structures of Gulf of Cambay from remote sensing. *International Journal of Remote Sensing*, 2169–2182.
Mohammad Hadian, A. M. (2021). Application of remote sensing technology in sediment estimating entering the dam reservoirs due to floods. *Shock and Vibration, 11*.
Narayan Swamy, G., & Sarma, R. V. (1982). Tidal mixing in dahej creek waters. *Mahasagar NIO Bulletin*, 195–203.
Nayak Shailesh R., & Sahai, B. (2007). Coastal morphology: A case study of the Gulf of Khambhat (Cambay). *International Journal of Remote Sensing*, 559–567. https://doi.org/10.1080/01431168508948478
Njenga, K. K. (2012). Velocity distributions and meander formation of river channels. *International Journal of Applied Science and Technology, 2*(9), 28–39.
Rouse, J. W., Haas, R. H., Schell, J. A., & Deering, D. W. (1973). Monitoring vegetation systems in the great plains with ERTS (Earth Resources Technology Satellite). In *Proceedings of 3rd Earth Resources Technology Satellite Symposium* (pp. 309–317). Greenbelt.
Rouse, J. W., Haas, R. H., Schell, J. A., & Deering, D. W. (1973). Monitoring vegetation systems in great plains with ERTS. In *Proceedings of the third ERTM Symposium* (pp. 309–317). US Government Printing Office, NASA
Shetye, S. R. (2011, May 11). Indian estuaries: Dynamics, ecosystems, and threats. *National Academic Science Letters, 34*(07 & 08), 229–237.
Singh, B. (2021). Evaluation of hydro-biological parameters in response to semi -diurnal tides in a tropical estuary. *Ecohydrology & Hydrobiology*. https://doi.org/10.1016/j.ecohyd.2021.03.002
Sinha, P. C., Jena, G. K., Jain, I., Rao, A. D., & Husain, M. L. (2010). Numerical modelling of tidal circulation and sediment transport in the Gulf of Khambhat and Narmada Estuary, West Coast of India. *Pertanika Journal of Science and Technology*, 293–304.
Slater, L. K. (2019). River channel conveyance capacity adjusts to modes of climate variability. *Scientific Reports, 9*(12619). https://doi.org/10.1038/s41598-019-48782-1
Srinivas, K., Das, V. K., & Kumar, P. K. (2005). Statistical modelling of monthly mean sea level at coastal tide guage stations along the Indian subcontinent. *Indian Journal of Marine Sciences*, 212–224.
SSNNL. (Updated July, 2022). *Narmada River and the Basin*. Retrieved from Sardar Sarovar Dam: https://www.sardarsarovardam.org/the-narmada-river-basin.aspx

Subash Bhandari, D. S. (2018). Landscape chande detection of Aliabet, Mouth Narmada River, Gujarat, using Geospatial technique. *JETIR*, 82–89.

Unnikrishnan, A. S., Shetye, S. R., & Michael, G. S. (1999). Tidal propagation in the Gulf of Khambhat, Bombay High and surrounding areas. *Proceedings of the Indian Academy of Science (Earth and Planetary Sciences)*, *108*, 155–177

Chapter 7
The Impact of Land Use and Land Cover Changes on Land Surface Temperature: Study of Udaipur Subdivision, Rajasthan

Khushbu Kulhari, Vinita Kumari, Priyanka, and Pooja Rani

Abstract Amid growing concerns about the impacts of climate change related events, this paper tries to establish pattern between related climate and climate-human induced coupled phenomena. The objectives of this study are to assess the land use and land cover (LULC) changes in the Udaipur administrative region of the state of Rajasthan (India). Using GIS-based remotely image-based data manipulation, this study substantiates and connects temporal changes (2000–2020) in the land-associated characteristics with land surface temperature (LST) changes in the region and net change in the vegetation extent using the Net Differentiated Vegetation Index (NDVI). Modelled geographically over an extracted Digital Elevation model, the region exhibits a significant decrease in dense forest cover by 31.4% over the period, while an increase in the cropland and built-up area is driven by the demand for infrastructural growth (in the forms of settlements, commercial estates, and industries) and the urban population clustering effect. In addition, the study found a negative relationship between the NDVI and LST variables, indicating a cooling influence of vegetation over the surface air band; however, the interpretation can further be improved (and/or substantiated) by incorporating the seasonal sub-soil moisture indices and precipitation. This study underscores the importance of remote sensing and GIS-based methods in monitoring socio-physical changes and positively guides policymakers in sustainable land-based management and practices. These findings are crucial in devising strategies for mitigating the long-term impacts, promoting climate resilient agricultural practices, enhancing urban planning, and fostering community-driven conservation efforts.

Keywords Land use/land cover · Climate change · Land surface temperature · Normalized difference vegetation index · Remote sensing and GIS

K. Kulhari (✉) · Priyanka · P. Rani
Department of Geography, Faculty of Earth Sciences, Mohanlal Sukhadia University, Udaipur, India
e-mail: kulhari101@gmail.com

V. Kumari
Department of Geography, University of Delhi, Delhi, India

© The Author(s), under exclusive license to Springer Nature Switzerland AG 2024
B. V. Raj Singh and A. K. Batar (eds.), *Sustainable Local Development for Environmental and Social Sustainability*, Human-Environment Interactions 11,
https://doi.org/10.1007/978-3-031-67303-0_7

7.1 Introduction

Climate change events are not new, the extent and frequency of extreme climate events are happening since the origin of this planet and these events are rising continuously throughout the world. Earth's climate is significantly changed by human activities at an unprecedented pace since the industrial revolution. Land surface temperature patterns help in understanding climate change by providing important information of environment. Anthropogenic activities linked to earth's surface transformations have been continuously causing environmental changes at local, regional, and global scales (Hua & Ping, 2018). Changing land use and land cover (LULC) due to the extension of agricultural areas, desertification, deforestation, and urban sprawl have significant influence on present and future environmental sustainability (Chaudhuri & Mishra, 2016). LULC change has been confirmed to have a significant impact on climate through various pathways that modulate surface temperature and precipitation (Feddema et al., 2005). Monitoring spatio-temporal changes in landscapes is thus becoming increasingly important in today's world which is facing challenges of increasing population, exploitation of resources, urbanization, and threats associated with climate change (Yue et al., 2007).

Advancement in remote sensing techniques, in terms of the availability of high-resolution spatial imageries over different periods of time and innovative image processing and data analyzing techniques, have led to efficient monitoring and investigation of dynamic changes in LULC (Chaudhuri & Mishra, 2016). Various studies have shown that LULC change detection provides useful information and better understanding of previous practices, current LULC, and future LULC (Almazroui et al., 2017; Bhagyanagar et al., 2012; Hussain & Karuppannan, 2021) and therefore helps in analysis of spatial extent and degree of human-induced and other environmental changes (Buyadi et al., 2013; Hua & Ping, 2018).

Land surface temperature (LST) retrieved from remotely sensed data is important parameter in assessing LULC changes and other characteristics of the earth surface (Kikon et al., 2016; Pal & Ziaul, 2017). There are many factors affecting LST such as soil content, moisture, change in vegetation cover, building material, and nature of surface area (Hussain & Karuppannan, 2021). Increased vegetation cover can decrease LST while increased built-up area and concretization of surfaces increase LST. Variations in LST can be characterized effectively using the values of NDVI. NDVI is a valuable tool which is used to monitor green cover growth, health of the vegetation, vegetation response to climate, and vegetation cover performance at landscape scale (Forkel et al., 2013; Hussain & Karuppannan, 2021; Zoungrana et al., 2018). Remote sensing and GIS have proved them very useful in accessing and analyzing LULC and LST, they have the ability to map and identify LULC and LST changes. This study uses these remote sensing techniques to assess urban change by calculating LULC for the years 2000, 2010, and 2020 and to calculate land surface temperature (LST) from 2000, 2010, and 2020 in the study region. Remote sensing techniques are used to examine vegetation change by calculating NDVI from 2000 to 2020 and the relationship between NDVI and LST from 2000 to 2020 in Udaipur subdivision.

7.2 Materials and Methods

7.2.1 Study Area

The study area comprises Udaipur Subdivision which includes six districts: Udaipur, Chittorgarh, Dungarpur, Pratapgarh, Rajsamand, and Banswara, located in Southern part of Rajasthan. The area of Udaipur subdivision lies between 23°03′N to 25°72′N latitude and 72°55′E to 75°11′E longitude. This region was part of the princely state of Mewar a few centuries ago, this region is also known as the Mewar region. Udaipur subdivision forms the inter-state boundary with the states of Madhya Pradesh and Gujarat. Udaipur subdivision has a total area of 36,634 km^2 which is 10.70% of total area of Rajasthan. This region comes under the world's oldest mountain range Aravalli where many peaks like Jarga, Kamalnath, Sajjangarh (Udaipur), and Kumbhalgarh (Rajsamand) are located. The region has diverse topography with plateaus, hills, and valleys, in the Eastern part of the region the rough hills are located and in the Northern part of this region the plain area is found due to the deposition of soil by the rivers and red loamy soil is predominant in this region. The presence of the Bhorat Plateau, Uparmal Plateau, and Bhomat Plateau in this region makes this region unique. In this area Banas, the largest interflowing river of Rajasthan flows, which further merges with the Chambal River and flows its water into the Bay of Bengal, whose major tributaries are Bedach, Kothari, and Gambhiri. The extension of the Mahi River is found in the southern part of this region, this region is known as Kanthal. Apart from this, Jakham, Som-Vakal, and Sabarmati Rivers are also found here and flow their water into the Arabian Sea. The Tropic of Cancer passes through the Banswara District located in this state, due to this, this region comes in both tropical and temperate zones. The average rainfall of this area is found to be 602 mm, and the amount of rainwater decreases from east to west and from southeast to northwest. The average temperature of January and May is found up to 17.13 and 34.36 °C (Fig. 7.1).

7.2.2 Landsat Data

Landsat-based images of 30 m spatial resolution covering the study area were collected for the years 2000, 2010, and 2020 from USGS, NASA's website. For identification of LULC, LST, Landsat images of Landsat 5 Thematic Mapper (TM), Landsat 8 Operational Land Imager (OLI) and Landsat 8 Thermal Infrared Sensor (TIRS), were used (Table 7.1). The LULC pattern in the study area is categorized

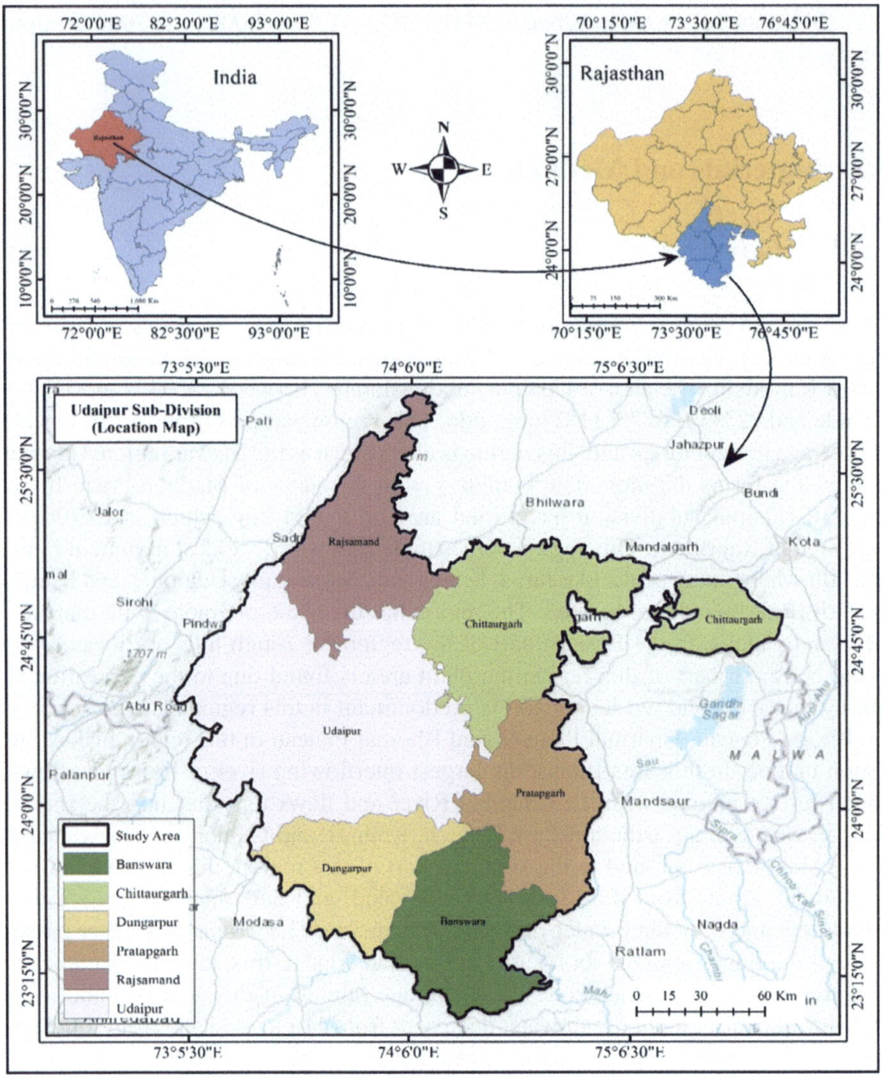

Fig. 7.1 Study area map of Udaipur subdivision. *Source* Gov. of India

into six classes: waterbodies, dense forest, grassland/crops, shrub land, built-up area, and barren land.

Table 7.1 Specification of Landsat satellite data

Different data source	Data used in the present study				
	Date	Resolution	No. of images	ROW	Path
Landsat 5TM	29th March 2000	30	5	147, 148	42, 43, 44
Landsat 5TM	25th March 2010	30	5	147, 148	42, 43, 44
Landsat 8 OLI	23rd March 2022	30	5	147, 148	42, 43, 44
Landsat 8 TIRS	23th March 2022	100	5	147, 148	42, 43, 44

Source USGS

7.2.3 Image Processing and Classification

Multispectral images from satellite were used for the analysis of LULC and LST. Extracted Landsat images were processed by using software ERDAS Imagine for layer stacking, mosaicking, and sub-setting of the image in accordance with study area (Rahman et al., 2017). After pre-processing, the supervised classification was applied on temporal datasets for the years 2000, 2010, and 2020. For each LULC type, training samples were selected by delimiting polygons using ArcGIS software (Nayak & Fulekar, 2017). Spectral signatures of different LULC classes were collected using pixels enclosed by these polygons. The detailed methodology is represented in Fig. 7.2.

7.2.4 Accuracy Assessment

Accuracy assessment was considered an important process in the valuation of several image processing in classification. Error matrix with information on the actual and predicted classification was the greatest means to present accuracy assessment. Overall accuracy is calculated by using the given formula (Hussain & Karuppannan, 2021) (Tables 7.2, 7.3 and 7.4):

$$\text{Overall Accuracy} = \frac{\text{number of sampling classes classified correctly}}{\text{Number of reference sampling classes}}. \tag{7.1}$$

Kappa values measure the accuracy of RS classification with the reference data. Kappa is calculated by the following method (Safder, 2019):

$$K = \frac{\text{Observed Accuracy} - \text{Chance Assessment}}{1 - \text{Chance Agreement}}. \tag{7.2}$$

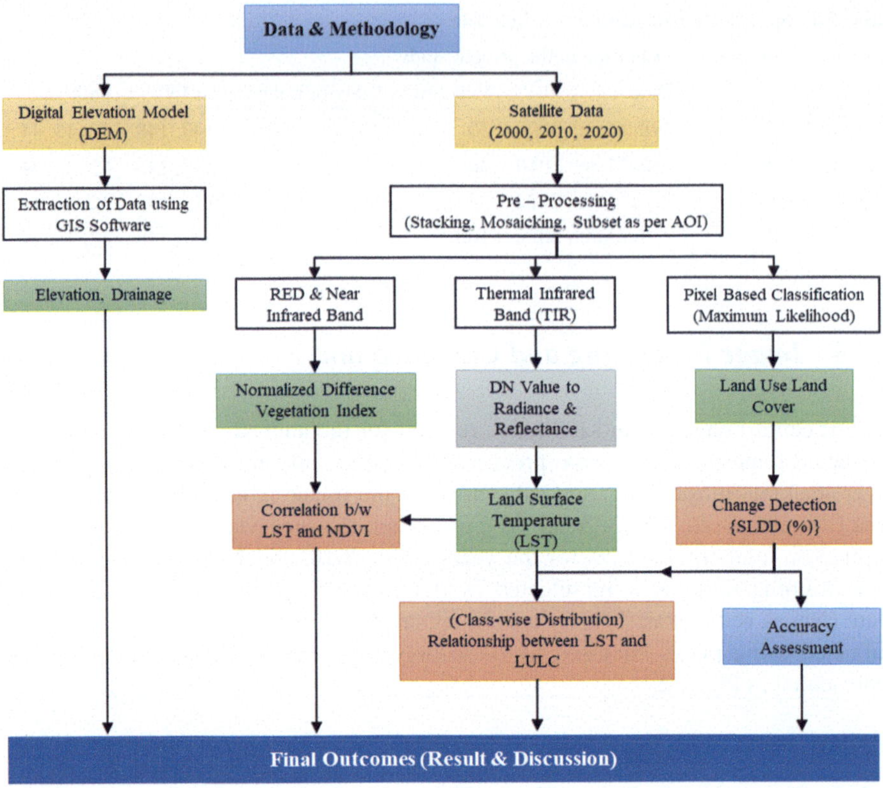

Fig. 7.2 Methodology flowchart

7.2.5 Estimation of NDVI

Normalized Vegetation Cover Index was calculated for monitoring change in vegetation. The NDVI values are calculated as follows (Mohammed et al., 2019):

$$\text{NDVI} = \frac{\text{NIR} - \text{RED}}{\text{NIR} + \text{RED}}, \tag{7.3}$$

where NIR is near-infrared band and RED is red band of satellite imagery.

7.2.6 Estimation of LST

The LST values were estimated from the thermal band of Landsat TM and OLI. Firstly, the $L\lambda$ values were estimated to spectral radiance by equation:

Table 7.2 Error matrix for different LULC classes for the year 2000

LULC	Water	Dense forest	Grassland/crops	Shrub land	Built-up area	Barren land	Total	UA
Water	23	1	1	2	1	0	28	82.14286
Dense forest	1	26	2	0	0	0	29	89.65517
Grassland/crops	1	2	27	1	2	0	33	81.81818
Shrub land	0	1	1	35	1	3	41	85.36585
Built-up area	2	1	1	1	29	2	36	80.55556
Barren land	0	2	0	0	1	30	33	90.90909
Total	27	33	32	39	34	35	200	
PA	85.18518519	78.78788	84.375	89.74359	85.29412	85.71429	DT→	170
OA	0.85	85		*K*	0.819565			

PA: producer accuracy; UA: user accuracy; OA: overall accuracy
Source Primary calibrations

Table 7.3 Error matrix for different LULC classes for the year 2010

LULC	Water	Dense forest	Grassland/ crops	Shrub land	Built-up area	Barren land	Total	UA
Water	30	1	1	0	1	0	33	90.90
Dense forest	2	25	1	0	0	1	29	86.20
Grassland/ crops	1	3	29	2	1	0	36	80.55
Shrub land	0	3	2	31	1	1	38	81.578
Built-up area	0	0	1	3	27	4	35	77.142
Barren land	1	0	1	0	1	26	29	89.65
Total	34	32	35	36	31	32	200	
PA	88.23	78.12	82.85	86.11	87.096	81.25	DT	168
OA	0.84	84		*k*	0.80785			

PA: producer accuracy; UA: user accuracy; OA: overall accuracy
Source Primary calibrations

Table 7.4 Error matrix for different LULC classes for the year 2020

LULC	Water	Dense Forest	Grassland/ Crops	Shrub land	Built-up area	Barren Land	total	UA
Water	26	1	1	0	0	0	28	92.86
Dense Forest	1	31	0	0	1	0	33	93.94
Grassland/ Crops	0	1	28	1	1	1	32	87.50
Shrub land	1	1	1	32	1	2	38	84.21
Built-up area	0	0	1	0	35	0	36	97.22
Barren Land	0	1	2	0	1	29	33	87.88
Total	28	35	33	33	39	32	200	
PA	92.86	88.57	84.85	96.97	89.74	90.63	DT→	181
OA	0.91	90.50		*k*	0.89			

PA: producer accuracy; UA: user accuracy; OA: overall accuracy
Source Primary calibrations

$$L\lambda = \text{gain} \times \text{QCAL} + \text{offset},$$

where $L\lambda$ represents spectral radiance and QCAL is quantized calibrated pixel value in DN.

Second step is to change spectral radiance value to temperature by equation:

$$T = \frac{K_2}{\ln\left(\frac{K_1}{L\lambda} + 1\right)}, \tag{7.4}$$

where K_1 and K_2 are 607.76 and 1260.56 for Landsat 4/5 (TM), and 772.88 and 1321.07 for Landsat 8 (OLI), respectively (Orimoloye et al., 2018).

Finally, the temperature in Kelvin scale was converted to Celsius using the following equation-

$$T(\circ C) = T(K) - 273.15$$

7.3 Result and Discussion

7.3.1 Land Use and Land Cover (LULC) Changes

The analysis of LULC changes using supervised classification technique from 2000 to 2020 showed that Udaipur subdivision was covered with different land use and land cover such as water bodies, dense forest, grassland/crops, shrub land, built-up area, and barren land as shown in Fig. 7.3. In the year 2000, dense forest covered 30.61% of area and grassland and shrub land covered 26.2 and 24.68%, respectively. Area under built-up was 3.08% while barren land was 11.78%. In the year 2010, area under grassland/crops was 38.08% followed by dense forests 26.37% and shrub land 24.7%. Area under barren land reduced to only 3.43% and water bodies also shrank to 2.73%.

Built-up area increased to 4.68% of total area. On the other hand, in year 2020, shrub land covered 35.40% and grassland/cropland area decreased to 34.2%. Water bodies were further reduced to 2.49%. The built-up area increased to 6.78% while barren land reduced merely to 0.14% of total area. The area under the built-up has seen an increasing trend from 2000 to 2020 in the study area. Results indicate that vegetation area and barren land were converted to roads, commercial buildings, and residential areas. Increasing population also arose need for expansion of agricultural land and more built-up area (Table 7.5).

7.3.2 Accuracy Assessment

Details of user's and producer's accuracy for different LULC categories for the year 2000, 2010, and 2020 are shown in Table 7.6. The highest and lowest producer's accuracy as well as user's accuracy values observed for 'built-up area' was 89.74% (2020), 85.29% (2000), and 97.22% (2020), 77.14% (2010), respectively. Similarly, for 'barren land' the highest and lowest producer's accuracy as well as user's accuracy

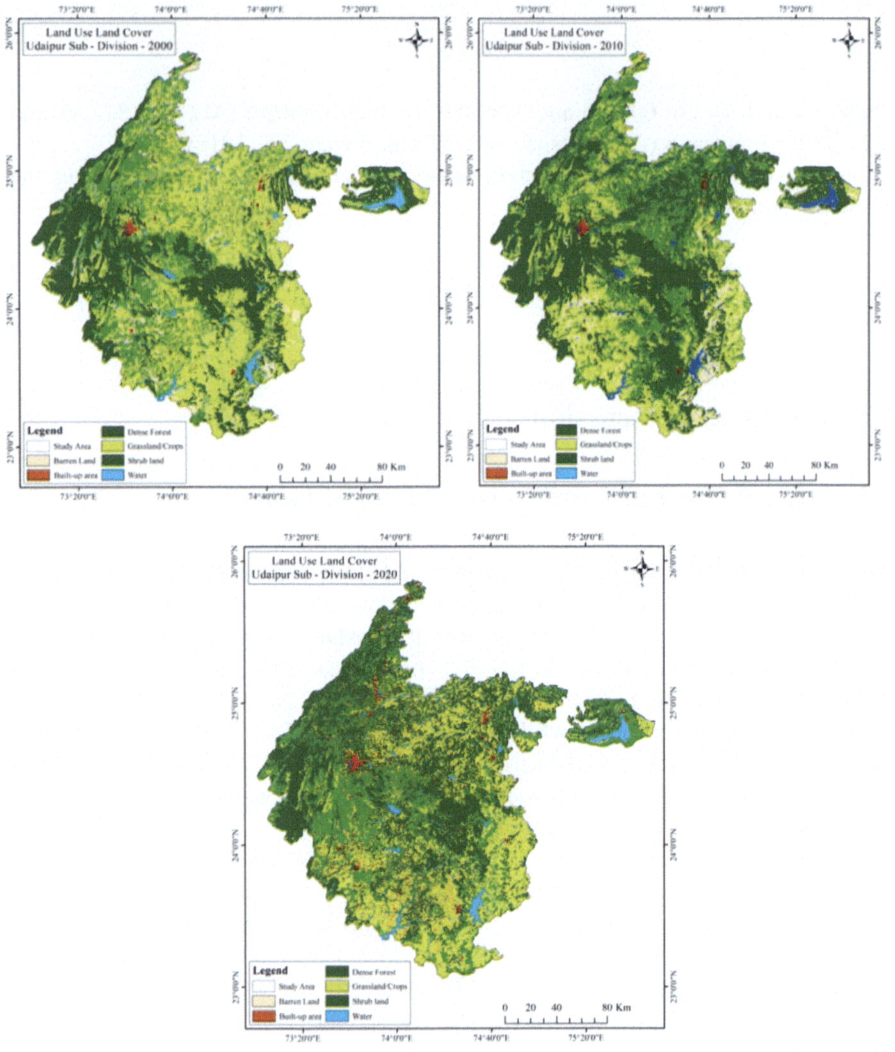

Fig. 7.3 LULC maps for the years 2000, 2010, and 2020 of Udaipur subdivision. *Source* Computed from USGS

observed was 90.63% (2020), 85.71% (2000), and 90.91% (2000), 87.88% (2020), respectively. Overall accuracy for the classification for the studied years is also reasonable, i.e., 85, 84, and 90% for the years 2000, 2010, and 2020, respectively. Kappa coefficient values were obtained 0.82, 0.81, and 0.89 for the years 2000, 2010, and 2020, respectively (Table 7.6). The producer's accuracy is the map accuracy, but the user's accuracy is point of view of map user (Hussain & Karuppannan, 2021). In the results, both the accuracies are in fair range, which is determinant of fewer errors in the classification.

Table 7.5 LULC areal distribution from 2000 to 2020 in Udaipur subdivision

Year	2000		2010		2020	
LULC class	Area (km^2)	%	Area (km^2)	%	Area (km^2)	%
Water	1355.70	3.65	1012.97	2.73	922.87	2.49
Dense forest	11,357.43	30.61	9784.66	26.37	7784.66	20.98
Grassland/crops	9720.84	26.20	14,130.6	38.08	12,690.3	34.20
Shrub land	9156.15	24.68	9166.3	24.70	13,136.4	35.40
Built-up area	1141.05	3.08	1736.31	4.68	2516.81	6.78
Barren land	4372.46	11.78	1272.79	3.43	52.59	0.14

Source Calculation based on USGS satellite data

Table 7.6 Summary of producer's and user's accuracy and kappa (K) coefficients in Udaipur subdivision

LULC	2000		2010		2020	
	PA	UA	PA	UA	PA	UA
Water	85.19	82.14	88.24	90.91	92.86	92.86
Dense forest	78.79	89.66	78.13	86.21	88.57	93.94
Grassland/crops	84.38	81.82	82.86	80.56	84.85	87.50
Shrub land	89.74	85.37	86.11	81.58	96.97	84.21
Built-up area	85.29	80.56	87.10	77.14	89.74	97.22
Barren land	85.71	90.91	81.25	89.66	90.63	87.88
Overall accuracy	85.00		84.00		90.50	
Kappa coefficient	0.82		0.81		0.89	

PA: producer accuracy; UA: user accuracy
Source Primary calibrations

7.3.3 Land Surface Temperature (LST) Changes

The spatial pattern and areal distribution of LST of three years (2000, 2010, and 2020) in Udaipur subdivision are represented in Fig. 7.4. The LST spatial patterns also highlight the changes in LULC classes. The range of LST was estimated 19.64–56.93 °C in 2000, 21.5–50.99 °C in 2010 and 23.44–42.64 °C in 2020 (Table 7.7). On an average, the mean temperature increased from 38.29 to 41.63 °C from 2000 to 2010 but then it again decreased to 35.95 °C in 2020. But there was a continuous increase in minimum temperature recorded over the region. The decrease in temperature can be attributed to increase in grassland/cropland and shrub land in the region from 2000 to 2020. High temperatures were recorded in Banswara and Pratapgarh districts and western Udaipur in 2000 but the area under high temperature decreased in 2010 due to increased area under dense forest and shrub lands. In 2020, due to increase in area of grassland crops, average temperature has decreased and zone of high temperature has also spread spatially in central part of the subdivision.

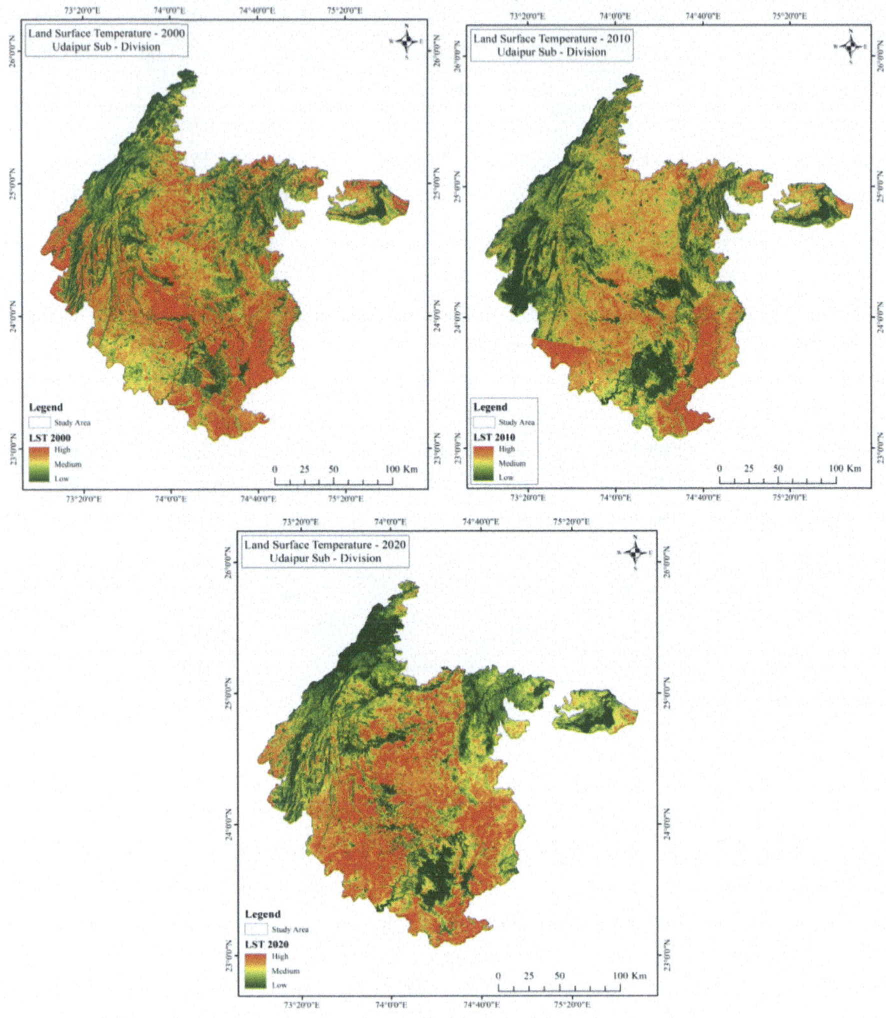

Fig. 7.4 LST maps for the years 2000, 2010, and 2020 of Udaipur subdivision. *Source* Computed from USGS

Table 7.7 Min, max, and mean values of LST and NDVI of the year 2000, 2010, and 2020

Year	Minimum		Maximum		Mean	
	NDVI	LST	NDVI	LST	NDVI	LST
2000	−0.92	19.64	0.97	56.93	0.06	38.29
2010	−0.54	21.5	0.72	50.99	0.037	41.63
2020	−0.24	23.44	0.48	42.64	0.135	35.95

Source Primary calibrations

7.3.4 Normalized Difference Vegetation Index (NDVI) Changes

Assessment of NDVI using Landsat image is based on the spectral properties of vegetation which reflects near-infrared (NIR) radiation. The NDVI model was derived using the Landsat images of 2000, 2010, and 2020 shown in Fig. 7.5. The NDVI values ranges were −0.92 to 0.97 in 2000, in 2010 the range was −0.54 to 0.72, and in 2020 the NDVI values changed to minimum −0.24 to maximum 0.48 (Table 7.7). Higher NDVI values show greater reflectance of infrared rays showing the most productive areas such as vegetation and forest. On the other hand, the lower values of NDVI represent the least or less productive areas such as barren land, built-up area, and water bodies. Overall the NDVI value is higher for dense and healthy vegetation. There is a large change from 2000 to 2010 due to increased shrub lands but then it decreased in 2020 where on small area shows higher NDVI value.

7.3.5 Relationship Between LST and NDVI

Correlation between NDVI and LST is shown in Fig. 7.6. For the year 2000, the regression line shows a negative relationship between NDVI and LST with a value R^2 0.06, and for the year 2020, the same negative correlation is observation with value of R^2 0.064. This shows that greater biomass of vegetation cover represents lower LST. Increase in vegetal cover can reduce LST. But LST is not only affected by NDVI, which can be seen by positive relationship between LST and NDVI in the year 2010. There are other factors such as rainfall and soil moisture which affect the LST. The explanation of this different relationship between LST and NDVI over different years needs further analysis over the study area.

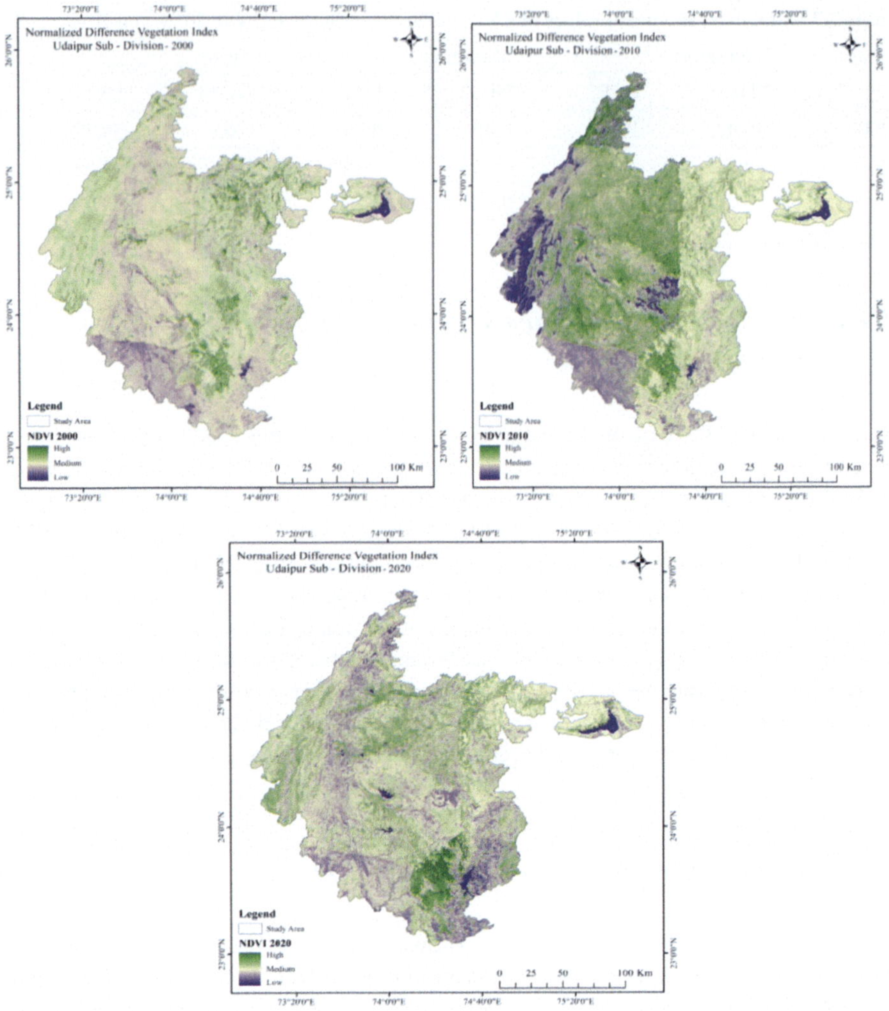

Fig. 7.5 NDVI maps for the years 2000, 2010, and 2020 of Udaipur subdivision. *Source* Calculated from USGS, NASA

7.3.6 Collective Actions Taken to Improve the Sustainability of Region

Local communities, NGO's, and government departments are working together to manage local common lands and forests and to reduce land degradation. This involves undertaking soil and water conservation work, protecting a site by constructing a boundary wall, planting trees suitable to the ecosystem and residents' requirements,

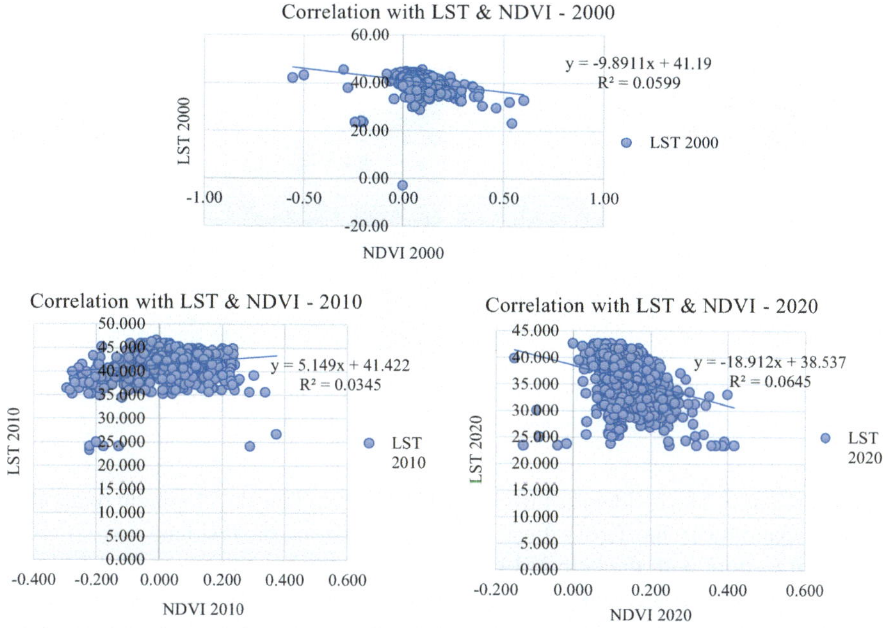

Fig. 7.6 Relationship between LST and NDVI for the years 2000, 2010, and 2020 of Udaipur subdivision. *Source* Primary calibrations

and putting in place management systems appropriate for the community. Some success stories are:

In 2002, villagers of Karech (a village in Udaipur) observed a decline in local green cover. Through collective campaigns aligned with the government's Joint Forest Management program, since 2002 the community has successfully restored more than 359 ha of common land. Strict rules were laid down to protect the young saplings so that they would not be grazed or cut, so the common grounds would remain closed for certain periods of the year; only dry, deadwood could be collected and not green, live trees, and so on. When communities enter the commons to gather cattle fodder, another rule is followed *ek ghar se ek datli* which means only one member of every household can collect fodder to prevent overuse (Fig. 7.7).

Over the years, the increased green cover has helped improve groundwater levels, thus allowing the soil to retain moisture. Karech has now become an exemplar of resilience and sustainability against climate change.

In Udaipur and Bhilwara, local communities are working with the Foundation for Ecological Security (FES), a non-profit that works with communities across the country on ecological restoration to manage common land. In 2016, residents of Boojh of Udaipur District formed a pasture development and management committee. The committee decided to revive 15 ha of pastureland by fencing the patch and planting 2500 saplings of local tree species. The next step was to build loose boulder check dams to harvest rainwater and to ensure that the hilly terrain

Fig. 7.7 Karech in 2002 and 2015. *Source* https://www.thebetterindia.com/257710/fighting-climate-change-rajasthan-villagers-turn-barren-land-to-dense-forest-him16/

retains moisture. The residents of Boojh in 2019 initiated the development of another pasture, spanning 16 ha. Tirol village of Udaipur District, the residents have revived a 25 ha of pasture with the help of FES and plan to revive another patch of 27 ha.

Collective actions were taken to reduce land degradation in Chitravas (a village in Udaipur). Physical intervention like rebuilding the stone wall was taken to protect the forest land. Planting of *jatropha curcas* cuttings along the stone wall was also done to provide an additional vegetative barrier. Several soil and water conservation measures including loose boulder check dams, gully plugs, contour trenches, and gabions to ensure in situ conservation of soil and moisture were constructed. Plantation of 50,000 saplings of various local species and sowing of grass seed was undertaken to improve the vegetation (Fig. 7.8).

Customs and traditions also help in conserving forests. The Kesar Chhanta Pratha is a nature conservation custom that involves the sprinkling of saffron on the name of the God Kesariya ji in the Udaipur and Dungarpur districts of Southern Rajasthan. The people of these areas believe that if they cut trees and forest, God will punish them. Thus, they take an oath not to cut the forest and preserve it for the future. In Piplantri (a village of Rajsamand), there is a custom to plant 111 saplings in the village every time a girl child is born.

Fig. 7.8 Forest management committee with locals performing rituals to 'protect trees' in Jhadol Phalasiya belt of Udaipur. *Source* Times of India (2020)

7.4 Conclusion

Based on the detailed analysis and findings presented in this study, it is evident that land use and land cover (LULC) changes have significant impact over the land surface temperature (LST) anomalies in the region. Over the period from 2000 to 2020, there have been marked differences in healthy vegetation growth especially in categories such as dense forest cover which decreased while categories such as the cropland and built-up area show a significant jump in the extent. These changes can largely be attributed to population growth and infrastructure development demand in the region, which leads a land cover change from natural vegetation and barren lands being brought under agricultural fold and settlement space. The study highlighted the critical role of remote sensing and GIS applications in monitoring and analyzing these dynamic changes. The correlation between LST and NDVI indices indicates a strong influence of vegetation cover on temperature regulation, with increased vegetation generally leading to lower LST. However, the complex interplay of factors such as soil moisture, rainfall, and human activities necessitates further detailed studies to fully understand the variability in LST. The insights gained from this study can aid policymakers, planners, and local communities in devising strategies to mitigate the adverse effects of LULC changes and balance development needs with ecological preservation. This includes promoting sustainable agricultural practices, enhancing urban planning to reduce heat islands, and fostering community-driven conservation efforts. In conclusion, the findings from this study provide a valuable framework for understanding the environmental impacts of land use changes, their relation between the socio-economic aspirations of the region and its people.

References

Almazroui, M., Masshat, A., Assiri, M., & Butt, M. (2017). Application of landsat data for urban growth monitoring in Jeddah. *Earth System Environment, 1*(2), 28. https://doi.org/10.1007/s41748-017-0028-4

Bhagyanagar, R., Kawal, B. M., Dwarakish, G. S., & Surathkal, S. (2012). Land use/land cover change and urban expansion during 1983–2008 in the coastal area of Dakshina Kannada district, South India. *Journal of Applied Remote Sensing, 6*(1), 3576. https://doi.org/10.1117/1.JRS.6.063576

Buyadi, S., Mohd, W., & Misni, A. (2013). Impact of land use changes on the surface temperature distribution of area surrounding the National Botanic Garden, Shah Alam. *Procedia-Social and Behavioral Sciences, 101*, 516–525. https://doi.org/10.1016/j.sbspro.2013.07.225

Chaudhuri, G., & Mishra, N. (2016). Spatio-temporal dynamics of land cover and land surface temperature in Ganges-Brahmaputra delta: A comparative analysis between India and Bangladesh. *Applied Geography, 68*, 68–83. https://doi.org/10.1016/j.apgeog.2016.01.002

Feddema, J., Oleson, K., Bonan, G., Mearns, L., Buja, L., & Meehl, G. (2005). The importance of land-cover change in simulating future climates. *Science, 12*, 1674–1678.

Forkel, M., Carvalhais, N., Verbesselt, J., Mahecha, M., Neigh, C., & Reichstein, M. (2013). Trend change detection in NDVI time series: Effects of inter-annual variability and methodology. *Remote Sensing, 5*(5), 2113–2144. https://doi.org/10.3390/rs5052113

Hua, A., & Ping, O. (2018). The influence of land-use/land-cover changes on land surface temperature: A case study of Kuala Lumpur metropolitan city. *European Journal of Remote Sensing, 51*(1), 1049–1069. https://doi.org/10.1080/22797254.2018.1542976

Hussain, S., & Karuppannan, S. (2021). Land use/land cover changes and their impact on land surface temperature using remote sensing technique in district Khanewal, Punjab Pakistan. *Geology, Ecology, and Landscapes, 7*, 46–58. https://doi.org/10.1080/24749508.2021.1923272

Kikon, N., Singh, P., Singh, S., & Vyas, A. (2016). Assessment of urban heat islands (UHI) of Noida City, India using multi-temporal satellite data. *Sustainable Cities and Society, 22*, 19–28. https://doi.org/10.1016/j.scs.2016.01.005

Mohammed, A. A., Shankar, K., & Hasan, R. N. (2019). Data on time series analysis of land surface temperature variation in response to vegetation indices in twelve Wereda of Ethiopia using mono window, split window algorithm and spectral radiance model. *Data in Brief, 27*, 4773. https://doi.org/10.1016/j.dib.2019.104773

Nayak, D. P., & Fulekar, M. H. (2017). Coastal geomorphological and land use and land cover study on some sites of Gulf of Kachchh, Gujarat, West Coast of India using multi-temporal remote sensing data. *International Journal of Advanced Remote Sensing and GIS, 6*(1), 2192–2203.

Orimoloye, I. R., Mazinyo, S. P., Nel, W., & Kalumba, A. M. (2018). Spatiotemporal monitoring of land surface temperature and estimated radiation using remote sensing: Human health implications for East London, South Africa. *Environmental Earth Sciences, 77*(3), 7252. https://doi.org/10.1007/s12665-018-7252-6

Pal, S., & Ziaul, S. (2017). Detection of land use and land cover change and land surface temperature in English Bazar urban centre. *The Egyptian Journal of Remote Sensing and Space Science, 20*(1), 125–145. https://doi.org/10.1016/j.ejrs.2016.11.003

Rahman, M. T., Tabassum, F., Rasheduzzaman, M., Saba, H., Sarkar, L., Ferdous, J., & Islam, A. Z. (2017). Temporal dynamics of land use/land cover change and its prediction using CA-ANN model for southwestern coastal Bangladesh. *Environmental Monitoring and Assessment, 189*(11), 565. https://doi.org/10.1007/s10661-017-6272-0

Safder, Q. (2019). Assessment of urbanization and urban sprawl analysis through remote sensing and GIS: A case study of Faisalabad, Punjab, Pakistan. *International Journal of Academic Research in Business and Social Sciences, 9*(4), 5811. https://doi.org/10.6007/IJARBSS/v9-i4/5811

Times of India. (2020, October 28). Tribal custom revives forests in Udaipur district. Times of India. https://timesofindia.indiatimes.com/city/udaipur/tribal-custom-revives-forests-in-udaipur-dist/articleshow/78900976.cms

Yue, W., Xu, J., Tan, W., & Xu, L. (2007). The relationship between land surface temperature and NDVI with remote sensing: Application to Shanghai Landsat 7 ETM + data. *International Journal of Remote Sensing, 28*(15), 3205–3226. https://doi.org/10.1080/01431160500306906
Zoungrana, B. J., Conrad, C., Thiel, M., Amekudzi, L. K., & Da, E. (2018). MODIS NDVI trends and fractional land cover change for improved assessments of vegetation degradation in Burkina Faso, West Africa. *Journal of Arid Environments, 153*, 66–75. https://doi.org/10.1016/j.jaridenv.2018.01.005

Chapter 8
Understanding the Discourses of Changing Urbanism from the Perspective of East Kolkata: A Study of the Wetlands

Sana Huque and Shailesh Yadav

Abstract Urban studies are no longer looking only into the spatial structure and function of a built landscape. Urban centres of today can be assumed to be synonymous with thriving entities with unique personalities moulded by their cultural, social, political and economic organisations. In this context, the idea of "urbanism" becomes important because it has been accepted to have become "a way of life" where there has occurred an "aestheticization of everyday life". In other words, cities are being planned in a manner that they represent an extension of the inhabitants' lifestyles. Ideas of sustainability and meticulous planning have become go-to phrases whilst envisioning such urban set-ups. But what warrants investigation is whether such plans are inclusive in terms of ecological, economical as well as social rights of every inhabitant. This paper attempts to explore the above concerns attached to the modern city set-ups with hybrid ecology where the natural and the man-made are often placed in juxtaposition. For a city like Kolkata, which holds dear the ambitions of becoming a global city, land as a means of promoting development is very difficult to come by. The recent expansion of the urban built-up area into the adjoining farmlands, water bodies and ecologically sensitive areas has been identified to be the reasons that are pushing the city towards ecological distress. The paper will use the case of East Kolkata Wetlands in order to question the extent of sustainability of the recent developments surrounding the East Kolkata Wetlands and look not only into the ecological perspectives but more importantly engage with issues pertaining to livelihood security and equity.

Keywords Ecology · Equity · Planning · Sustainability · Urbanism · East Kolkata Wetlands

S. Huque
Nature Conservation Foundation, Bengaluru, India

S. Yadav (✉)
Department of Higher Education, Jaipur, Rajasthan, India
e-mail: geoshailesh@gmail.com

© The Author(s), under exclusive license to Springer Nature Switzerland AG 2024

B. V. Raj Singh and A. K. Batar (eds.), *Sustainable Local Development for Environmental and Social Sustainability*, Human-Environment Interactions 11,
https://doi.org/10.1007/978-3-031-67303-0_8

8.1 Introduction

The study of urban landscapes over the years has transcended from its traditional focus on spatial structure and functions of built landscapes to include diverse range of multidisciplinary issues. Urban centres of today are therefore defined by their cultural, social, political and economic organisations. Moreover, as the population numbers the world over are shifting in favour of urban areas, there has been growing acknowledgement that cities need to be built differently. These factors have played an important role in addressing the ecological as well as the social sustainability of cities. The focus of the paper is to explore the relationships that exist in hybrid ecology where the natural and the man-made are often placed in juxtaposition. The emphasis shall be to explore the city's understandings of urbanism and how accommodative it is to not just sustainable but also equitable living. The present paper has been drawn from a recent study of the city of Kolkata. The objective of the study broadly lies in the city's increasing conflicts arising out of its newfound aspirations.

It needs to be mentioned that a study of Kolkata becomes important because in recent years the city and its periphery have undergone significant changes. Dasgupta (2007) writes in "*City Divided?*" that this emerging "new Calcutta" (present Kolkata) with its burgeoning private sector under active encouragement of the government and the planning and development agencies of the state is all set to redesign the spatial parameters of the city, especially the fringe area. Such developments are taking place by glossing over marginalisation of the already marginalised communities. Therefore, the space for the poor in Kolkata has been shrinking by the moment. The East Kolkata Wetlands, that continue to shelter several families who continue to make a living off the land, forms the crux of this study, and it focuses on analysing how the city is adapting itself to fit into the idea of development and what consequences it pertains for different category of inhabitants.

8.2 Understanding the Discourses of Changing Urbanism in Contemporary India: What Do the Theories Reflect?

Urban studies have seen a dynamic evolution. The early theories like the Concentric Zone Theory[1] or the later schools of Urban Studies like the Los Angeles School[2] and the New York School[3] laid the foundation on which current understandings have been based. However, these aforementioned models, although have had immense contribution in understanding the development of cities the world over, continue to be rooted in western understandings. In the context of South Asia in general and India in particular, attempts to fit these models have proved to be exercises similar

---

[1] Proposed by Robert E. Park and Ernest W. Burgess in the 1920s that stated that urban spaces get compartmentalised into zones due to competition for land and other resources.

[2] Rejected the domination of urban core and put greater emphasis on the hinterlands.

[3] It took the focus back to the city center similar to the Concentric Zone Model.

to fitting "square pegs in round holes". Roy (2005) points out how there exists a duality in the study of urbanisation between the First World Global Cities and the Third World Megacities. For the purpose of studying the megacities of the south, it is pertinent that indigenous models and theories are developed. However, till the time such developments take place, Roy directs towards the route of policy responses in order to explain the developments taking place in the southern cities. In Roy's suggested context, it needs to be mentioned here that the adoption of policies which resulted in the transition of India from socialist to neoliberal has had a major impact on not only the economy of the nation but also how the cities of the country shall be perceived in the future in terms of social, ecological and cultural perspectives.

It needs to be noted that the early ideas in urban studies themselves contained hints of conflicts between communities or between the city centre and the hinterlands. But such conflicts have become prominent urban phenomena over the years that were to follow. In developing nations, the conflicts deepened further since they were hit by the wave of neoliberalism in the 1990s. The current trend that has picked up in almost every urban centre as a consequence of popularisation of neoliberal ides is the tendency amongst the middle class and the upper class to evoke discourses of "public interest" and "citizenship" to table civic concerns that are in reality garbs to push the exclusion of the poor living in urban areas (Baviskar, 2011). This tendency has been termed by Baviskar (2011) as "*Bourgeois Environmentalism*", and Fernendes (2004) terms it as "*Politics of Forgetting*". Urban planning also in its attempt to market cities for new industries and markets has strongly imbibed this tendency. So, the present cities in general are aspiring primarily to be "economic space of highways, market areas, and commuter zones" (Campbell, 1996). Under such circumstances, Wirth's (1938) "urbanism as a way of life" and Knox's (2011) "aestheticization of everyday life" seem to have been taken to be understood in the literal sense. In today's planned cities, therefore the Lefebvrian idea of "Right to the City"[4] (Lefebvre, 2003; Harvey, 2003) is on its way out with the space for the poor on a rapid decline and capitalist and state agencies are working in tandem for production of exclusive arenas for the urban elite.

Almost simultaneously with the spread of neoliberalism, in the 1990s, another major change was strengthening its footing on the global fora. It was the concept of *sustainable development* which had begun to leave its mark on developmental ideas. The idea, "*Development that meets the needs of the present without compromising the ability of the future generations to meet their own needs*", mentioned in the Brundtland Report of 1987 was fast becoming the foundation over which all human activities were beginning to be vetted to determine their compatibility with the environment. However, there is more to the idea of sustainable development than its popular environmental component. The idea also advocates the need for achieving social and economic equality which are often forgotten when planning for sustainable growth.

[4] A slogan coined by Lefebvre in 1968 to signify the right to access urban resources and more importantly have the right to make the city one's own by making changes at the individual level which David Harvey called as an important human right.

The triple ideals—ecological, economical and sustainable—that summarise the idea of sustainable development have also become closely intertwined with what is political over time. Over the world, this blend has given a positive voice to those who have been systematically marginalised. But in cities of developing nations, as understood from the works of Baviskar, Campbell and Fernandes, co-mingling of these ideas has been used more often than not as an advantage by the privileged to further marginalise those who already inhabit the periphery.

As early as 1944 Karl Polanyi had discussed in *The Great Transformation* how the spurt in capitalist market can prove detrimental to the society[5] as well as the environment (O'Connor, 1986; Smith, 2006). In the current situation, the economic system has also closely intertwined itself with political forces. Therefore, the impacts felt on the environment and society have been greatly influenced by the politics of the region also. The following section is a detailed discussion of the situation where the nature and political landscape come together to create a framework for the urban way of life.

8.3 The Nature–Politics Nexus: Understanding the Political Ecology Framework in the Context of Changing Urbanism

As mentioned in the previous paragraphs, politics and nature have become closely intertwined over the years. The works of Piers Blaikie in Political Ecology helped understand that environmental changes have significant social and political impacts (Forsyth, 2008), particularly on those who are classified as poor or the vulnerable in society. Political Ecology in other words has helped to academically acknowledge that the complex relation nature and society share with each other determines "access and control over resources and their implications for environmental health and sustainable livelihoods" (Watts, 2000). Therefore, in the context of our study, it is necessary to put under scrutiny questions like "whose nature is, or becomes, urbanized and the uneven power relations through which this socio-metabolic transformation takes place" and then closely examine the politics through which "nature becomes urbanized and contemporary cities are produced [which are] both enabling and disempowering dwelling spaces for humans and non-humans alike" (Heynen et al., 2006).

The restructured global capitalism has had noted impact on the politics of environment and politics of distribution (Watts, 2000). In the context of cities, they have taken the form of urban environmentalism. Brand and Thomas (2005) say that "urban environmentalism…can be seen as forming a constitutive part of one of the greatest transformations of space in urban history. It has been hugely successful, not in its own overt reference to ecology, but in the sense of providing argumentative and

[5] Adverse impact because of increase in uneven capitalist development, class exploitation and other forms of struggles.

representational support for the rebuilding of cities, the spatial restructuring of urban economics, new forms and techniques of government, urban lifestyles, a sense of citizenship and political subjectivities".

The above explanations have given a backdrop against which the wetlands of Kolkata have been observed. The case of Kolkata, with the tussles over its public space and ecological assets, fits the bill of the above understandings about environmentalism and politics in urban settings. In Kolkata, every prominent political party which rose in popular imagination by mobilising the poor to fight for their cause later turned in favour of the well-to-do classes once having found a prominent footing in the political scene of the city. This is why anti-poor exercises like *Operation Sunshine* was hailed as "the world's worst city...cleaning up its act" and "both official and popular discourses cast the event as a return to *bhadralok* Calcutta" (Roy, 2011a, b). In similar instances of the state flip-flopping between the poor and the privileged, it first pushed the formal city to adopt informal means in a manner that rules and norms of governance appear to be capricious and arbitrary (Roy, 2011a, b). Using practices like "vesting" the state managed to convert agricultural land into settlement colonies for sharecroppers, squatters and resettled slums in the 1970s and 1980s. Now with the increase in demand for land for development, the state government through agencies like West Bengal Housing Board and Kolkata Metropolitan Development Authority has involved itself in promoting townships like Rajarhat New Town, the East Kolkata Township and the Baishnabghata-Patuli Township in the same areas by displacing the previous poor occupants. The wetlands of East Kolkata are witnessing such acts on a widespread scale where the fisher folks and the agriculturalists are losing out to the big development projects that are coming into the area.

8.4 Background of the Study Area: Why Wetlands of East Kolkata?

The study of the field area, i.e. the wetlands of the East Kolkata, requires a vivid description from various socio-ecological contexts. Wetlands are known the world over for their high productivity. A study by the Millennium Ecosystem Assessment (MEA) in the year 1997 estimated that the wetlands meet almost 45% of the planet's productivity in spite of covering only about 7% of the Earth's surface. The East Kolkata Wetlands are important from this perspective because it provides a wide array of goods and services, including ecological services, to the city of Kolkata.

Most significantly, the wetlands of Kolkata are significant for its paddy production involving low cost of cultivation. The wetlands have been used to cultivate paddy by using sewage water, thus setting an example of successful resource recovery system. In similar use of sewage wastes of the city, the wetlands have successfully made sewage fed fishing into an important economic activity.

The wetlands are important for ecological services like carbon sequestration too. The city needs to realise that for a thriving city, it is a great boon to have a carbon

Table 8.1 Some goods and services from East Kolkata Wetlands and the loss from conversion between 1990 and 2000

Goods and services	Loss from conversion
Paddy cultivation	Rs. 87 million p. a
Fish production	Rs. 14.15 million p. a
Waste recycling and resource recovery	Rs.0. 87 million p. a

Source Chattopadhyay (2001)

pool at its proximity. The wetlands are helping the city from choking on its own wastes and pollutants. Hence, they are an important microclimate stabiliser as well.

In terms of biodiversity too, the wetlands hold value. The East Kolkata Wetlands' biodiversity spectrum ranges from microorganisms to carnivores. The wetlands are home to several kinds of reptilian, avian and mammalian populations. However, the conversion of the wetlands for residential and institutional purposes has been a major reason for eroding them along with the many benefits they provide (Chattopadhyay 2001). The table below outlines some of the losses suffered as a result of conversion of the East Kolkata Wetlands (Table 8.1).

Because of the above-mentioned functions, the East Kolkata Wetlands have been accorded the recognition of being a unique ecosystem that is critical for environmental sustainability of not just the state of West Bengal but for the entire subcontinental watershed (Banerjee, 2012). In spite of the ecological value of the area and the apparent governmental efforts in protecting them, the wetlands continue to be violated by the expanding urban landscape of the city of Kolkata. It has become apparent that the rate of urban expansion into the wetlands, which attracted the government's attention for its conservation in the 1980s, has accelerated further since the 1990s.

The wetlands actually cover vast stretches extending from the north-east, east to the south-east of the city extending into the threshold of the Sundarbans. According to Banerjee (2012), this area derives its distinct ecological characteristics from the low-lying spill basin that has been the consequence of the tidal activity of the river system comprising of Bidyadhari-Piali-Matla that have been in turn responsible for embedding the wetland ecosystem in the broader ecosystem of Ganga–Brahmaputra delta. As a consequence, the region is considered to be biological and geomorphic continuum of the estuarine Sundarbans that lie to the south and the east of the wetlands.

The area began to be considered an integral part of the city of Kolkata since the late nineteenth century or early twentieth century onwards. Although the adverse human impacts on the wetlands began to be felt since the time of the Colonial occupiers, it was not until the 1990s that the threats to the wetlands have aggravated. Even the massive wave of migrants from East Pakistan (now Bangladesh) who came in the 1950s and 1960s and settled in the eastern periphery of the city on land claimed from the wetlands did not prove to be as disastrous for the fragile ecosystem as the recent developments in the area (Banerjee, 2012) (Fig. 8.1).

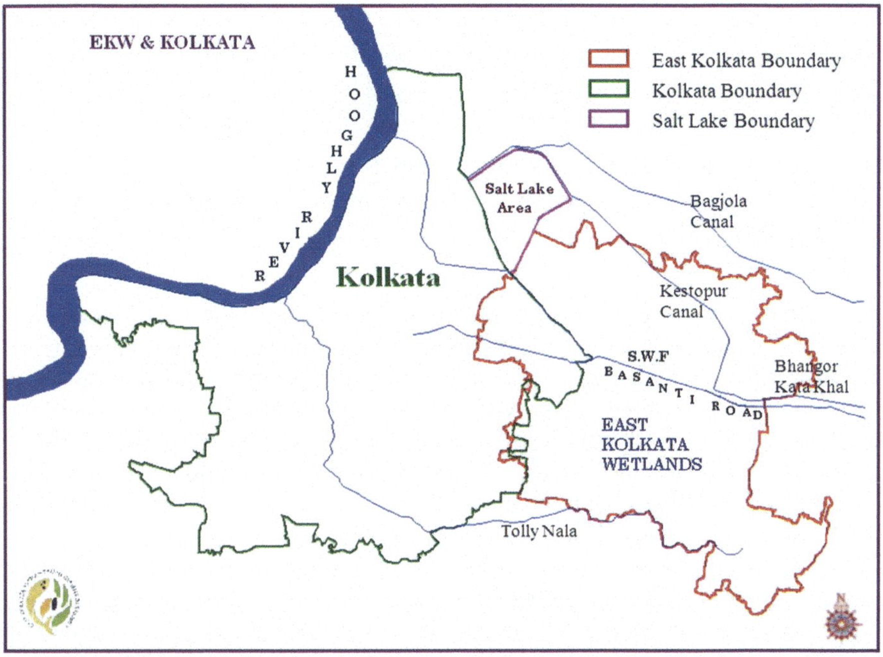

Fig. 8.1 Map of the East Kolkata Wetlands. *Source* EastKolkataWetlandsManagementAuthority. Accessed August 02, 2024

8.5 Methodology

Whilst studying the area for its socio-ecological importance, the emphasis was placed on ethnographic approaches that involved unstructured interviews, participatory observation method, and focus group discussions. The study was conducted in the month of June, 2014 which stretched into the following month of August. Households that were surveyed for the study were in Kumarpukur Gram of Kheyadah mouza of Sonarpur Police Station. Official figures state that 20,000 families continue to reside within the wetland complex ("East Kolkata Wetlands", 2010, p.3). Out of this, the area is inhabited by more than 3000 tribal families belonging to tribal groups of Mundas, Santhals, Sardars and Oraon. The area is notable for its high incidence of poverty despite the fact that the area itself is notified for being highly resource rich. East Kolkata Wetlands Management Authority itself acknowledges that 77% of the families residing in the area are under the poverty line as the average household income is 70% less than the state average ("East Kolkata Wetlands", 2010).

For the purpose of household survey, the sampling method used was random sampling for the respondents. About 30 households were covered in this exercise. It was observed that very few houses in the area have access to potable drinking water or adequate sanitation facilities. Infrastructure facilities like electricity, access to hospital and institutions for education is also poor in the area which has several tribal households. From the conservation perspective, the area holds value.

8.6 Using the Political Ecology Framework in Understanding the Development of Wetlands in East Kolkata

The incidence of poverty is remarkable in the area. This contrasts greatly with the richness of resources that this area is endowed with. The act of benevolence by the state of settling poor families in the wetlands area has had very limited impact on alleviation of poverty that was observed to be widespread. Political parties, especially those belonging to the CPI(M), played a proactive role post land reforms in ensuring that the less privileged have access to land as a source of livelihood. However, the political activists' crusades for the poor did not go beyond resettlement. No further action was taken to consolidate the rights of the poor over the land.

The rising real estate demands in the areas in and around the wetlands have the families staring at an unpredictable future. There has been a rise in livelihood loss and displacement of the families as the demand for land to develop has gone up over time. The problem has been aggravated due to the fact that most of the families settled are extremely poor and do not own any *pattas* or ownership documents that would help them prove that they are the ones who are engaged in tilling the lands and running the fish farms. In terms of people's resistance movements the area has seen very few effective ones. However, the recent developments leave a great scope for political dissent and struggle for the rights of those affected by the conversion of the wetlands as an extension of the city's built-up area.

Roy's (2011a, b) writings in *Re-forming the Megacity* help to understand that the densely populated slums in Calcutta do not represent the limits of the city's growth. Rather the emerging enclaves at the rural edge of the city have become the frontiers of the expanding city. Roy (2011a, b) observes that these frontiers are being increasingly marked by dissent and struggle because planning and governance, that are the instruments facilitating territorial expansion and legitimization, are called into question by the emerging social movements and political alliances. Thus, it can be said that urban politics and not land use which determines the logic of planning in a city like Kolkata. The trend of planned townships and enclaves of development in a city that suffers from massive infrastructure deficiencies have resulted in remarkable urbanization in formerly rural expanses at the periphery of the city. The following observation highlights two important perspectives that present the issues and arguments about the development of the New Township in Rajarhat.

The first perspective discussed is that of Mitra (2002), the erstwhile Managing Director of HIDCO, who wrote that the New Town developed in Kolkata can be presented to the country as an alternative model of urban development that involves upgrading infrastructure by the state government. The model, according to the author, goes beyond existing perception involved with such developmental activities where mostly the poor section of the society is left to fend for their survival in the envisaged plan of minimised participation of the state and unfettered liberty for the market to operate. According to Mitra, the Rajarhat development project was a smooth amalgamation of actions of an "activist" state and efficiency of market where the poor were the biggest winners.

On the other hand, one of the citizen platforms of the city, Nagarik Mancha, claims that in the garb of development of Rajarhat, actions like displacement and environmental degradation were carried out under the sponsorship and legitimising approval of the state machinery (De, 2012).

Mitra (2002) pointed out in his article that like most planned urbanisation, Rajarhat's development also occurred in the periphery of the "existing urban or metropolitan agglomeration". He outlined certain objectives that were proposed as the reason for developing a new town for Kolkata:

- Help minimise the urban pressure on the main metropolitan area;
- Promote planned urban growth;
- Create a new Central Business District for the city and
- Increase the number of housing stock, especially for the economically weaker section of the city.

The objectives for developing Rajarhat outline very well the immediate need for developing an alternative for Kolkata's metropolitan zone that seems to be bursting at the seams. However, Nagarik Mancha has come up with observations that negate the above-mentioned objectives, particularly the last one. Nagarik Mancha raises the question as to why the *Urban Land (Ceiling & Regulation) Act, 1976* was done away with. The organisation sees that such an action by the state government, at the behest of World Bank, was to facilitate the removal of the poor occupiers of land and allow the middle class and the upper class to have more land available for them to occupy.

Mitra (2002) contended in his article that the development of Rajarhat New Town has been able to prove erroneous that planned urbanisation in India is characterised by land acquisition problems, lack of participation of the local, pervasive speculative and anti-poor deals and "capture" by the middle and upper classes. He gave examples of planned townships from India as well as abroad to prove how they have failed to address concerns of the poor and hence the overall sustainability of such projects and contrasts Kolkata's New Town project. In the words of the Mitra, Rajarhat New Town has been notable for:

- Strong pro-poor approach;
- Recognition of the state as an enabler and facilitator rather than a doer;
- Intensive involvement of the local land losers;
- Considered anti-speculative measures;
- Low budgetary impact and
- Specific, targeted subsidies.

Nagarik Mancha however points out that local people who owned agricultural plots in the area have been protesting against land acquisition for the new town right from the moment the project was conceived. "Rajarhat Jomi Bachao Committee" or "Committee to Save Rajarhat's Land" is proof of people's unwillingness to give up their land for the development project.

The paper by Mitra (2002) also claims that most of the land identified for the New Town lies in the zone of smallholdings, poor irrigation intensity and low agricultural productivity. He also says that the area's agricultural operations were impacted because the area had a tendency of getting water logged while ingress of brackish water into the area at periodic intervals further aggravated the problems. In addition to the above-mentioned issues, the areas adjacent were experiencing a burst of unplanned urban growth. As a consequence, the local residents were under relentless pressure to sell their land to facilitate real estate growth. In such a condition, speculators and middlemen had begun to operate with impunity. According to Mitra (2002), all these factors were facilitators in mobilising favourable public opinion and also accelerating land assembly.

Again Nagarik Mancha's narrations of development of Rajarhat appear to be very different from Mitra's observations. It states that the CPI(M) enjoyed a political monopoly in the area and the party's actions were further facilitated by the fact that it was also the political party that had been in power in the state for a very long time. As a consequence, the political party faced no major resistance when bulldozers were let loose on agricultural fields or tonnes of soil were dumped on standing crops that were ready for harvest. Even pumps from the irrigation department were made defunct without prior notice at the time of paddy transplantation and in some instances sand was offloaded on agricultural fields in order to render the soil infertile for future agricultural activities. The people affected at that time did put up a resistance. However, without any political or administrative support, their movement fizzled out.

Nagarik Mancha states that till 2010 the land acquired for the township was about 2421 hectares out of which 91% was farm land. As can be expected, majority of those affected were marginal farmers. Other facts that need to be stated are that in 2000–2001, out of 3958 hectares of farmland in the area, 60% of them were suitable for multi-cropping. Loss of agricultural crop productivity was not the only concern. The area also lost several hectares of fisheries whose fish production were over 10,000 quintals (Table 8.2).

Apart from a loss of livelihoods, the developmental activity also led to the loss of biological assets whose impacts shall be felt over the years to come. The West Bengal

Table 8.2 Loss in annual paddy production due to conversion of EKW

Police station	Area of land (ha)	Quantity (quintal)
Sonarpur	388	14,744
Rajarhat	600	33,000
Tiljala	250	9500
Total	1238	57,244

Source Chattopadhyay (2001)

Government's Fisheries Department itself acknowledged that the Rajarhat area has a unique combination of diverse geographical and natural features. 80% of the area has wetlands. It is also the habitat of various floral and faunal assets. In addition, the area has several fallow riverbeds, mud areas, saline wetland, lowland marshes, ponds, large water bodies, canals which have ecological significance but have been lost because of the developmental project.

8.7 Replicating the Poor and the Planned Development Paradox in East Kolkata Wetlands

The developments taking place in and around the East Kolkata Wetlands remarkably appear to be reflecting the paradox the poor face in a planned urban setting. The land which until recently was looked upon as unfit for elite consumption has now become prime location for development of property worth millions. The concerns about such developments have been how rapidly they have managed to shrink the space available to the poor. As discussed previously in the paper, the wetlands have since several generations been the home and source of livelihood for several tribal people and people who continue to make a living through primary activities. The sudden spurt in planned urban development that only has space for gated residential towers, luxurious malls and office space has the poor staring at a future that involves loss of their livelihoods and displacement. Therefore, the question arises that if the development of the city in general and the wetlands in particular is meant primarily for the IT workers and global citizens, then "what room remains for the millions of Kolkatans who do not fit into this particular vision of the future?" (Bose, 2013). What the planning agencies as well as the state need to realise is such developments are neither desirable nor sustainable. The paradox is reinforced when the state takes on itself the role of determining who can lay claim over the space it governs.

Scott (1998) writes in his book that towns or cities that were built in the times of middle age or medieval times have a particular look—"it is the look of disorder". In other words, the urban centres did not conform to any plan or abstract imagination for it. This lack of a planned development was not a matter of concern for those living in them. However, this chaotic arrangement that has historically been synonymous with demand for greater political autonomy amongst the masses became a matter of concern for authorities. Policing and control by the state was found to be facilitated by promotion of spatial intelligibility that could be attained only through planning (Scott, 1998). In the current world, planned cities not only enable the state to fulfil its role as a monitor but also empower it to determine who can stake a claim to its space. The result has been the prominent privileged classes aggressively cornering the available space and excluding those who do not possess the ability to wield similar power.

Planned cities and poverty have a paradoxical existence because of existence of above-mentioned situations—the greater the planning the more aggravation of penury. However, it must be kept in mind that it is difficult to pin the extent of positive or negative impact of planning on development as it varies greatly across cities. Planning should be seen as just one amongst the range of complex and interlinked factors that play a role in determining the extent of poverty and unsustainable development. Therefore, it can be said that it is unlikely that planning is the dominant factor influencing unsustainability and poverty in the cities. Nevertheless, literature is available in significant numbers that suggest that urban planning particularly in global south has not succeeded in addressing the primary urban issues of present times. Moreover, some argue that faulty planning is the reason behind worsening poverty and environment in the cities. The arguments that are used to support such claims basically are focussed on two aspects which are as follows: firstly, ideas borrowed from planning experiences from the global North are imposed which are ill-suited in the context of cities in the developing world, and secondly, there is misuse or abuse of planning systems in context of local politics, and manipulation of factors which are economic or cultural. The trend today is that of urban modernism. The following are some of the major characteristics of urban modernism as outlined by Watson (2009):

- Major emphasis is placed on the aesthetic appearance of the cities. The visualisation of such spaces is that of being organised, expansive, efficient, green and clean with grand views of the city's state and civic buildings. Such spaces, to live up to the standards, therefore have very low tolerance of poor people and informal activities.
- Implementation of the "superblock concept" which stands for high-rise buildings that have low plot coverage and large setbacks. This leaves large stretches of open spaces between concrete structures.
- Planning for free-flowing vehicular movement routes as the assumption is that car ownerships in cities are high.
- The land use functions are planned to be segregated using zoning regulations. Therefore, the city is segregated into areas for residence, community facilities, commerce, retail and industry.
- Most importantly, residences segregated according to the different groups.

Although the idea of urban modernism appears to be appealing superficially, delving deeper into the concept reveals that it is not capable of accommodating the way of life for a majority of the urban population. For the rapidly growing poor cities that are dominated by informal settlements and occupation structures, urban modernism leads to social as well as spatial marginalisation.

Modernisation projects breed social segregation and inequality because of their inherent assumptions. Promotion of monofunctional areas like residential segregation further adds to economic burden of the poor as in many cases the dwelling units also double up as economic units for the low-income households (Feitosa et al., 2008). Moreover, modernist cities hardly provide any space to carry out informal business. Thus, it is not only a form of social exclusion but it is also denying of livelihood to the poor as the high-end localities are also areas which give opportunity to access the purchasing power of the higher income market. In the case of the East Kolkata

Wetlands, the paradox has become prominent because of the juxtaposition of elite localities within areas that still have visible hutments, paddy fields and fish ponds.

8.8 Some Possible Ways Forward

The paper, attempts to bring out the fact that a city's desire to mould itself to suit a neoliberal world does not just lead to drastic changes for the city alone; it transforms natural and ecologically sensitive areas as well. This renders natural commons inaccessible to the least advantaged groups. The following paragraphs attempt to suggest a few measures that can assist in taking forwards the discourse of conservation of the wetland area of Kolkata without alienating the farmers and fishers who continue to reside and make a living from it.

Recent understandings of natural resource management indicate that local communities need to be included in the processes of conservation. Appreciation of local communities as custodians of natural resources has become an accepted idea. Therefore, a conservation plan for the wetlands that involves the local community and gives scope for the representation of their needs and aspirations can be expected to bring positive results in conserving the fragile area.

Cooperative conservation is another method that can be popularised in the wetlands. This conservation scheme not only involves the local communities but also has the scope to bring together government agencies, industries, as well as the people of the city. The approach also believes that the immediate convincing factor for the local communities to save the wetlands can be through financial incentives. Also, the strengthening of property rights of the local residents is expected to be effective against coercion to vacate land for real estate. However, this approach extends the incentive-induced approach a little further by emphasising upon the development of market-based action plans that would make conservation of the area the preferred economic option amongst those who own, manage or derive their livelihoods from the wetland. For instance, as per this approach, landowners can be paid compensation for conserving land and providing environmental services that the public wants and benefits from (The Cooperative Conservation Blueprint, n.d.). An ecosystem services market incentives plan can be formulated that can monetise the ecosystem services that the area provides to the city. So services like flood control, water purification and recharge, and carbon sequestration can become paid services that public or private utilities can avail by paying for them to wetland managers or landowners. This would result in creation of new sources of income for landowners and most importantly, protection and conservation of natural areas (Creating a Cooperative Blueprint, 2010).

Finally, the last suggestion is less of a measure that can be implemented on ground to address the problem and more of an academic exercise which can shed better light upon the problems plaguing areas like the East Kolkata Wetlands. This approach is to engage meaningfully with the concept of political ecology. Political ecology concept is not just about the predominantly discussed macro-understanding of effects of neoliberalism and the actions and policies of the state on the natural environment. The concept also has a side that focuses on the efforts and struggles of the socially disadvantaged groups in protecting the environment. Political ecology acknowledges the nuances present in the struggles for livelihood amongst different social actors and takes into account the cultural idioms that may vary in context due to local politics. Political ecology pays attention to social actors in urban settings (Heynes et al., 2006; Meyers, 1999) and their connection with degradation process (Moore, 1993). The unequal power relations and the resultant problems need to be identified before they can be dismantled and Political Ecology allows the understanding of such patterns of human–environment relation.

8.9 Conclusion

The wetland of East Kolkata, a valued environmental asset, has become the bone of contention between the environmental activists, political individuals and the real estate development sector. In this struggle, the poor and their cause appear to have got lost somewhere. As the city prepares to launch itself in the big league of urban development, the city has become more and more intolerant towards those who are deemed unfit to be living in modern cities. The cities have become the breeding ground for the environmentalism of the privileged that have proved time and again to be parochial and exclusivist. Samaddar's (2011) words very well reflect the aspirations that the city aims to achieve and gives a hint of the price that the city must pay for its aspirations:

> Kolkata has changed quite a lot in the last few decades. It wants to become Delhi. It must catch up with the flash and glitz found elsewhere. It too must have its high-tech township and must embody a new mode of circulation of money, information, human resources, and power. It does not think that its old organic character is worth retaining. If discarding the old organic character is the necessary sacrifice to make in order to develop, let that be. If road space increases while space for human interaction decreases that price Kolkata must pay.

Sammadar's laments if looked through the prism of current developments in the city appear to be prophesying the impending doom for the poor in the city. This is because the city has woken up to the realisation that it is time it undergoes a massive image haul over. Therefore, it is not just the city but also the fringe areas of Kolkata that are undergoing a remarkable transformation, particularly the eastern fringes of the city. The transformations are not just in terms of landscape but also in the social lives of the people who inhabit the fringes and eke out a living from activities like agriculture and fish farming. The loss of livelihood for many has been a devastating loss. There have been significant changes in social, political, economic as well as ecological aspects of the eastern fringe, and the dichotomy is more marked there than the rest of the city. The changes have been catastrophic for not just the ecologically sensitive wetlands but maybe for the entire city at large. It needs to be realised that the wetlands are crucial for the well-being of the future city of Kolkata. Indiscriminate destruction of the wetlands in the name of development may soon see the city facing a major ecological disaster. Development hence should strive to become more inclusive with ample space for both the human, irrespective of class, caste and category, and the non-human to thrive.

References

Banerjee, S. (2012). *The March of the Mega-city: Governance in West Bengal and the Wetlands to the East of Kolkata*. Humbolt University.

Baviskar, A., & Ray, R. (Eds.). (2011). *The elite and everyday: The cultural politics of the Indian middle classes*. Routledge.

Baviskar, A. (2011). Cows, cars and cycle-rickshaws: Bourgeois environmentalists and the battle for Delhi's streets. In A. Baviskar & R. Ray (Eds.), *Elite and everyman* (pp. 341–418). Routledge.

Bhan, G. (2009). This is no longer the city I once knew. Evictions, the urban poor and the right to the city in millennial Delhi. *Environment and Urbanization, 21*(1), 127–142.

Bose, P. S. (2013). Bourgeois environmentalism, leftist development and neoliberal urbanism in the City of Joy. In T. R. Samara, S. He, & G. Chen (Eds.), *Locating right to the city in the global south* (pp. 127–151). Routledge.

Brand, P., & Thomas, M. J. (2005). *Urban environmentalism*. Routledge.

Campbell, S. (1996). Green cities, growing cities, just cities? Urban planning and the contradictions of sustainable development. *Journal of the American Planning Association, 12*, 308–326.

Campbell, S. D. (2013). Sustainable development and social justice: Conflicting urgencies and the search for common ground in urban and regional planning. *Michigan Journal of Sustainability, 1*, 1122.

Cases On Protection Of Lakes. (n.d.). http://www.cseindia.org/node/2557

Chakravorty, S. (2005). From colonial city to globalizing city? The far from complete spatial transformation of Calcutta. In N. Fyfe & J. Kenny (Eds.), *The urban geography reader* (pp. 84–92). Routledge.

Chattopadhyay, K. (2001). *Environmental conservation and valuation of east Calcutta wetlands (wetland and biodiversity EERC working paper series: WB-2)*. Integrated Research and Action for Development. http://irade.org/eerc/pdf/WB_FR_KunalChattopadhyay.pdf

Creating a Cooperative Conservation Blueprint for Florida. (2010). Florida: Florida fish and wildlife conservation commission. http://conservation.dcp.ufl.edu/CCB%20Report.pdf

Dasgupta, K. (2007). A city divided? Planning and urban sprawl in the eastern fringes of Cacutta. In A. Shaw (Ed.), *Indian cities in transition* (pp. 314–340). Orient Longman Private Limited.

De, P. (2012). *Rajarhat: When the government loots*. Retrieved October 14, 2014, from http://sanhati.com/articles/5354/.

EKW & Kolkata. (n.d.). *East Kolkata Wetlands management authority*. https://ekwma.in/ek/maps-2/

East Kolkata Wetlands (2010, November). *The Newsletter of East Kolkata Wetlands Management Authority and Wetlands International-South Asia, 1*. https://www.ekwma.in/ek/wp-content/uploads/2015/09/NewsletterVolume1.pdf

Feitosa, F. F., Reyes, J., & Zesk, W.B. (2008). Spatial patterns of residential segregation: A generative model. Paper presented at X Brazilian Symposium on GeoInformatics, Geoinfo 2008, Rio de Janeiro.

Fernandes, L. (2004). The politics of forgetting: Class politics, state power and the restructuring of urban space in India. *Urban Studies, 41*(12), 2415–2430.

Forsyth, T. (2008). Political ecology and the epistemology of social justice. *Geoforum, 39*, 756–764. https://doi.org/10.1016/j.geoforum.2006.12.005

Harvey, D. (2003). The right to the city. *International Journal of Urban and Regional Research, 27*(4), 939–941.

Heynen, N.C., Kaika, M., & Swyngedouw, E. (Eds.). (2006). *In the nature of cities: Urban political ecology and the politics of urban metabolism. Volume 3 of questioning cities series.* Taylor & Francis.
Heynes, N., Perkins, H. A., & Ray, P. (2006). The political economy in race and ethnicity in producing environment inequality in Milwaukee. *Urban Affairs Review, 42*(1), 3–25.
Knox, P. (2011). *Cities and design.* Routledge.
Lefebvre, H. (2003). *The urban revolution.* University of Minneapolis.
Meyers, G. A. (1999). Political ecology and urbanization: Zanzibar's construction materials industry. *The Journal of Modern African Studies, 37*(1), 85–108.
Mitra, S. (2002). Planned urbanisation through public participation. *Economic and Political Weekly, 348*(11), 1048–1054.
Moore, D. S. (1993). Contesting terrain in Zimbabwe's eastern highlands: Political ecology, ethnography and peasant resources struggles. *Economic Geography, 69*(4), 380–401.
O'Connor, J. (1986). Capitalism, nature, socialism: A theoretical introduction. *Capitalism Nature Socialism, 1*, 11–38.
Roy, A. (2011a). *Calcutta requiem: Gender and the politics of poverty.* Pearson Education India.
Roy, A. (2005). Urban informality: Towards an epistemology of planning. *Journal of the American Planning Association, 71*(2), 147–158.
Roy, A. (2011b). Re-forming the megacity: Calcutta and the rural-urban interface. In A. Sorensen & J. Okata (Eds.), *Megacities: Urban form, governance, and sustainability* (pp. 93–109). Springer.
Samaddar, R. (2011). *Rajarhat, the Urban Dystopia.* http://transitlabour.asia/blogs/dystopia
Scott, J. C. (1998). Cities, people, and language. In *Seeing like a state* (pp. 53–83). Yale University Press.
Smith, N. (2006). Foreword. In Heynen, N.C., Kaika, M., & Swyngedouw, E. (Eds.), *In the nature of cities: Urban political ecology and the politics of urban metabolism. Volume 3 of questioning cities series* (p. xiv). Taylor & Francis.
Watson, V. (2009). The planned city sweeps the poor away...: Urban planning and 21st century urbanisation. *Progress in Planning*, 72(3), 151–193.
Watts, M. (2000). Political ecology. In E. Sheppard, & T. Barnes (Eds.), *A companion to economic geography.* Blackwell.
Wirth, L. (1938). Urbanism as a way of life. *American Journal of Sociology, 44*(1), 1–24.

Chapter 9
Detection of LULC Change with Erosion and Accretion in the Catchment Area of Panchet Reservoir, India, in the Context of Sustainable Development and Management

Ravi Mishra and Ishan Kumar Paliwal

Abstract Detection of land use–land cover (LULC) change based on remote sensing data is an important source of information for various decision support systems. Detection of land use and land cover changes provides a lot of information that is crucial for land conservation, sustainable development, and management of water resources. Therefore, the objective of this study is to find out the changes in land use and land cover in the area around Panchet Dam. The Panchet Dam, located on the margins of the states of West Bengal and Jharkhand, was constructed in 1959 by disrupting the Damodar River, which currently affects the vegetation around the catchment area. The main objective of this paper is to identify the changes in the LULC pattern and also to understand the erosion and accretion in the water area of the dam. Remote sensing data, satellite imagery, and image processing techniques have been used to identify land cover changes. The research was done on the basis of images that were taken in 1990, 2000, 2010, and 2020 using Landsat TM +30 m resolution. Erdas Imagine and ArcGIS software were used to identify the transformations. Seven land cover classes (water body, bushland, grassland, forestland, cultivated, and residential land) were taken in the classification. This study indicates that over the last 40 years, agricultural land and residential areas in the basin have increased by 15.61 and 8.05%, respectively. Therefore, appropriate land management practices, integrated watershed management, and active participation of the local community should be advanced to protect against unwanted LULC changes in the region.

R. Mishra (✉)
Ex-Student of Department of Remote Sensing and GIS, SSJ Campus - Almora, Kumaun University, Almora, India
e-mail: ravimishra5may@gmail.com

I. K. Paliwal
Department of Geography, Faculty of Earth Sciences, Mohanlal Sukhadia University, Udaipur, India

© The Author(s), under exclusive license to Springer Nature Switzerland AG 2024
B. V. Raj Singh and A. K. Batar (eds.), *Sustainable Local Development for Environmental and Social Sustainability*, Human-Environment Interactions 11,
https://doi.org/10.1007/978-3-031-67303-0_9

Keywords LULCC · Erosion · Accretion · GIS · Dam · Land management

9.1 Introduction

Land use and land cover changes play an important role in shaping the relationship between economic development and biodiversity conservation globally. The increase in the pursuit of economic development during the second half of the twentieth century resulted in the conversion of vast surface of earth, i.e., ~24% into the cropland and swiping away of ~35% mangroves and ~20% coral reefs. There has been an exponential increase on an average 3.9% annual growth rate in the global gross domestic period over the same period. But there was a possibility that 'aggregate capital stock' have declined because the profits generated from exploiting the natural resources are generally not invested in alternative forms of capital; rather, they are consumed.

The change in land use and land cover of an area is the result of natural and socioeconomic aspects as well as their operation by humans in time and space (Bajocco et al., 2012). Land use–land cover (LULC) changes are mostly influenced by population growth and reduction in an area, economic development, and many other physical factors including topography, slope conditions, soil types, and climate (Dewan & Yamaguchi, 2009). Land use change is a matter of historical process that deals with how people use land. It modifies the availability of various resources including vegetation, soil, and water. Land use–land cover change has become an important issue considering global dynamics and their responses to environmental and socioeconomic drivers. Land use–land cover changes negatively affect climate patterns, natural hazards, and socioeconomic dynamics at the global and local levels (Tewabe & Fentahun, 2020). Information about land use/cover and their best use prospects is essential for understanding the selection, planning, sustainable land resource management, and changes in hydrological processes to meet the increasing human needs and demands.

The Rio Declaration of 1992 clearly states that to achieve the goal of sustainability at the local level, the most important aspect is 'good governance' and therefore not only an efficient government but also sincere local stakeholders become a prerequisite. The major reason behind this argument is that the idea of sustainable development in its own is of a very high magnitude which cannot be achieved by the efforts of government alone. It is really important to understand the role of local communities, local government, and other stakeholders and bring them under an umbrella and channelize their efforts with the government at the upper level.

Detection of change is important in many applications related to land use and land cover (LULC), including farming, urban expansion, and landscape change (Ishtiaque et al., 2017). Understanding landscape patterns, changes, and interactions between human activities and natural phenomena is essential for proper land management and decision-making (Parker et al., 2003). Geographic information systems (GIS) and remote sensing (RS) are powerful and cost-effective tools to assess the spatial and temporal variation of LULCs (Baynard, 2013). Nowadays, remote sensing data and GIS are applicable and valuable for land use and cover change detection studies. Remote sensing data are the most common source for LULC pattern detection, while GIS is useful for quantification, and mapping, as it is suitable for repetitive data acquisition, and processing, and provides accurate georeferencing (Kibret et al., 2016).

A lot of research has been done on LULC change, which shows that LULC change studies are very important for environmental protection strategy and sustainable resource management of basins (Yesuph & Dagnew, 2019). Many regions in India are currently facing massive LULC transformation. Much of the area is undergoing changes from natural resources to agricultural systems and human settlements. Lakes or other water basins play a major role in detecting this change, as lull changes have a direct impact on these areas (Rashid & Aneaus, 2020). Thus, monitoring and assessment of land use dynamics require timely monitoring of LULC dynamics and quantification of landscape patterns for Panchet Dam.

Therefore, this study aims to assess the land use trend and land cover change in the study area. In this study, the LULC changes were investigated using remote sensing data with the help of ArcGIS and Erdas Imagine software. Therefore, we hope that the findings of this study can contribute to already available information for decision-makers, land managers, and planners for the sustainable development and management of natural resources.

9.2 Study Area

The study area of the Panchet Pahari Dam lies on the margins of the states of West Bengal and Jharkhand, India. The latitudinal and longitudinal extent of the region is 86°29′ East to 86°46′ East and 23°30′ North to 23°46′ N (Bid, 2016) (Bid & Siddique, 2019). It is surrounded by Purulia District in the east, Bankura District, and Barddhaman District in the south and in the northeast, and the north and northwestern parts of the Dhanbad district of Jharkhand. Maithon Reservoir is also located in the northeastern part of the region. The slope of the entire region is from west to east and southeast, marked by the direction of the currents flowing through the area. Damodar is the main river of this region. The region is characterized by a dry and humid sub-humid tropical climate with a southwest monsoon regime. According to Köppen, this region belongs to the ‘O’ type of climate. The temperature ranges from 3.8 °C in winter to 52 °C in summer, and the annual rainfall ranges from 1100 to 1500 mm. The soil in this area is generally of red laterite residual type. The existence of hills and undulating land are other features of the region. The region is covered with a wide variety of vegetation of the Sal, the Palas (Butea frondosa), and the tropical dry range (Kumar & Saikia, 2020). To the northeast of this region is the 643.5 m high Panchkot or Panchet hill, made up of Upper Gondwana sediments. Panchet Hill Dam is built 5 km above the confluence of Barakar at Damodar which is the tri-junction of Dhanbad, Barddhaman, and Purulia districts (Bid, 2016) (Bid & Siddique, 2020).

The position of the study area is shown in Fig. 9.1.

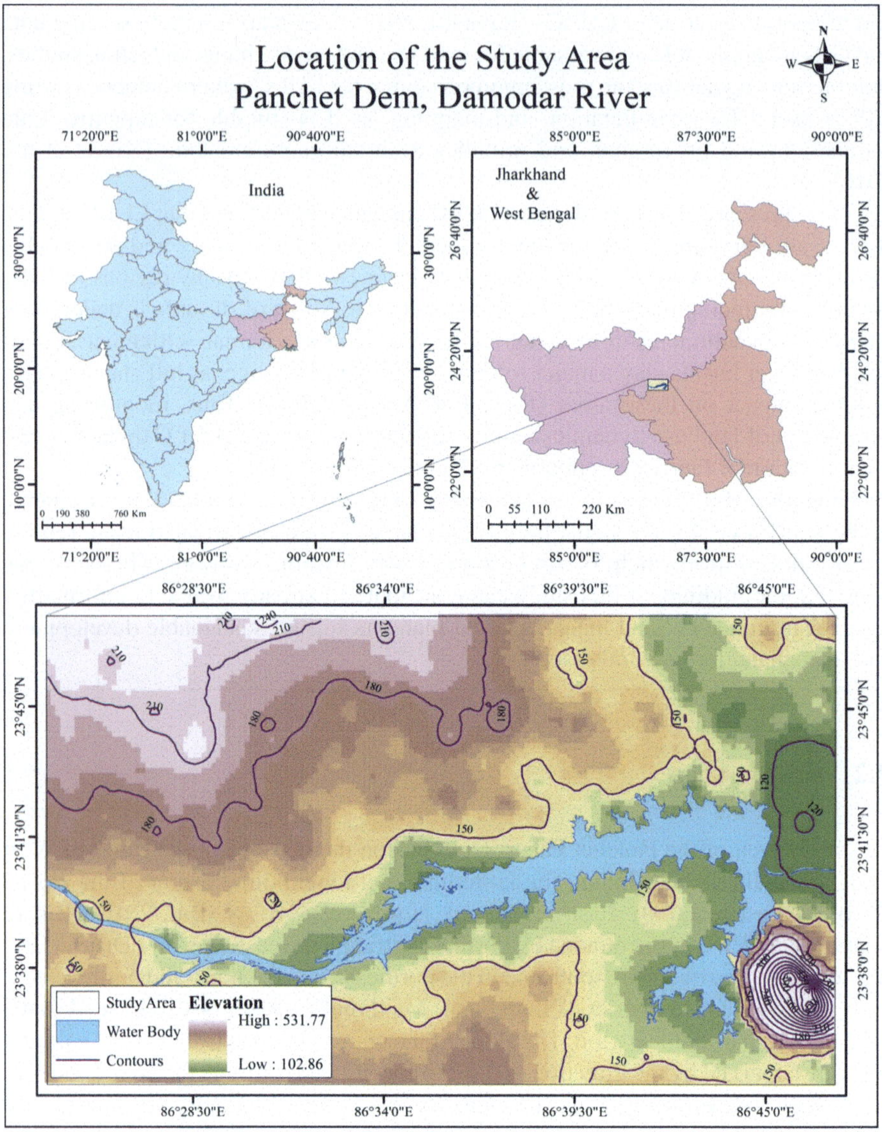

Fig. 9.1 Panchet Dam and its location

9.3 Database and Methodology

9.3.1 *Dataset*

Landsat image is a specialized sensor on the Landsat satellite platform and is capable of capturing multispectral imagery. Landsat multispectral images capture multiple electromagnetic spectrum bands, including visible, reflective, and thermal infrared energies (Short, 1982). Understanding the strengths and weaknesses of different types of sensor data is essential to the selection of suitable remote sensing data for image classification. The selection of appropriate sensor data is the first critical step for a successful classification and takes into account the user requirement, scale, and characteristics of the study area (Lu & Weng, 2007). Factors such as the availability of various image data and their characteristics, cost and time constraints, and the analyst's experience in using the selected image need to be considered. Change detection was analyzed in this study using LULC, and satellite images, employed in conjunction with ArcGIS and Erdas Imagine for automatic classification. Landsat satellite images of the region were obtained for the period 1990, 2000, 2010, and 2020. Therefore, this study is of interest in the detection of LULC changes in these periods using Landsat images. Landsat data are one of the most important and accurate datasets for understanding the global land cover status, provided independently by the United States Geological Survey (USGS). Images from 1990, 2000, and 2010 were used by Landsat 5TM (Thematic Mapper) and 2020 Landsat 8 OLI (Operational Land Imager). ASTER DEM data of 30 m resolution were used to prepare the elevation map of the area. Data were processed using Erdas Imagine and ArcGIS software. The topographic map is georeferenced by the Erdas Imagine 15 software and generates a shapefile to demarcate the catchment area of the Panchet Dam. Standard false-color composites (FCCs) are created from satellite images by band combining. The study area is extracted from standard FCC through ArcGIS software subsets.

9.3.2 *Methodology*

9.3.2.1 Land Use and Land Cover Classification

To perform image classification, the raw data must be preprocessed and properly prepared so that errors due to Earth's geometry, and radiometric and atmospheric effects can be accounted for. Common procedures in the preprocessing stage include detection and restoration of bad lines, geometric correction or image registration, radiometric calibration, atmospheric correction, and topographic correction (Anderson, 1976).

Supervised Image Classification

In this process, we select pixels that represent patterns we recognize, or that we can identify with the help of other sources (Google Earth) information: knowledge of the data, the class desired, and the algorithm to be used as required before starting the selection of training samples (Ma et al., 2017).

9.3.2.2 Assessment of Accuracy

A graded image or change detection map should be compared with reference data, which is considered correct, to assess its performance and determine its accuracy (Foody, 2002). This procedure was used to estimate the accuracy of image classification by comparing the classified map with a reference map. Therefore, a complete accuracy assessment needs to include reports on overall accuracy, user accuracy, and manufacturer accuracy, which were examined using the kappa coefficient (Foody, 2002).

9.3.2.3 Image Processing

The preprocessing phase included detection and restoration of bad lines, geometric correction or image registration, radiometric calibration, atmospheric correction, and topographic correction. Precise geometric correction or image registration of remotely sensed data is a prerequisite for combining different source data in a classification process. The study used Landsat imagery to identify changes in the LULC distribution in the basin over 30 periods from the years 1990–2020. To avoid seasonal variation and to reduce cloud cover, the dates of the obtained data were selected in the dry season and with the same annual season as the acquired years. To see and discriminate surface, input satellites are created using the RGB color structure. In addition, the images provided complete coverage of the study area, and mosaicing, stacking of bands, and sub-setting of the images were performed using the Erdas Imagine software.

9.3.2.4 Change Detection

The importance of change detection is to determine which land use class is changing to the other. The most commonly used land change detection methods include image overlay, classification comparisons of land cover statistics, change vector analysis, principal component analysis, image rationing and the differencing of land use–land cover. In this study, classification comparisons of land cover statistics were used. The areas covered by each land cover type for the various periods were compared (Lu et al., 2004). Then, the directions of the changes (positive or negative) in each land cover type were determined.

9.3.2.5 Erosion and Accretion

'Accretion' is the term applied to the gradual increase or acquisition of land by the action of natural forces like washing sand, soil, or silt from a watercourse or seashore. In contrast to accretion, 'erosion' is the gradual washing away of land along the shoreline (Bird, 1983) (Shukla & Gedam, 2019).

$$\text{Erosion} = \text{Area of Previous Year} - \text{Unchanged Area}$$
$$\text{Accretion} = \text{Area of Next Year} - \text{Unchanged Area}$$

9.4 Result and Discussion

9.4.1 Land Use and Land Cover Analysis

To assess classification accuracy, a confusion matrix including overall accuracy, user's accuracy, and manufacturer's accuracy was used. Classification is not complete until its accuracy is assessed using known kappa statistics (Tewabe & Fentahun, 2020).

9.4.1.1 Accuracy Assessment

To assess the accuracy of the LULC result, it has been compared to high-resolution Google Earth images and points of field investigation. For each class, 250 random and digitized reference points have been taken. The accuracy level is calculated based on the reference data. Google Image and field validation are shown in Table 9.1. The level of accuracy of the LULC results is presented and explained below.

9.4.1.2 Overall Accuracy

This accuracy gives the overall result of the confusion matrix. It is calculated by dividing the total number of correct pixels (the diagonal value) by the total number of pixels in the confusion matrix. According to Anderson (1976), the minimum accuracy value for reliable land cover classification is 85%. On the other hand, levels of accuracy are accepted by users and may not be acceptable to other users for a certain task.

Table 9.1 Summary of accuracy assessment from 1990 to 2020 (%)

Years	1990		2000		2010		2020	
LULC	PA	UA	PA	UA	PA	UA	PA	UA
Water bodies	92.50	90.24	90.63	85.29	88.89	91.43	96.77	93.75
Built-up area	86.49	86.49	82.35	90.32	81.25	86.67	88.57	93.94
Barren land	87.88	87.88	85.71	80.00	85.00	80.95	92.11	89.74
Fellow land	87.18	89.47	89.13	89.13	89.19	86.84	94.59	89.74
Grass land	89.74	94.59	84.78	88.64	88.89	88.89	92.31	94.74
Shrub land	90.63	85.29	93.75	93.75	89.19	89.19	94.44	91.89
Dense forest	90.00	90.00	90.63	87.88	90.63	90.63	91.18	96.88
Overall accuracy	89.20		88.00		87.60		92.80	
Kappa coefficient	87.38		85.93		85.52		91.59	

9.4.1.3 Kappa Coefficient (KC)

The kappa statistic is used to measure the agreement between two sets of classifications of a dataset (Table 9.1) and is used to estimate the accuracy of the predictive model by measuring the agreement between the predictive model and a set of field surveyed sample points (Moriasi et al., 2007). Accuracy has been performed differently for 1990, 2000, 2010, and 2020 images, respectively, using Google Earth.

The results showed that the 1990 overall accuracy was 89.20% while the manufacturer's and user's accuracy ranged from 86.49 to 92.5% and 85.29 to 94.59%, respectively. The kappa statistics were calculated from the error matrix, and the coefficient of classification for 1990 was 87.38%. Thus, according to the classification scale given by Moriasi et al. (2007), the classification ranks very well.

Generally, the results (Table 9.1) show that the overall accuracy for the years 2000, 2010, and 2020 was 88, 87.60, and 92.80%, respectively. This result met the minimum accuracy (Anderson, 1976) for land cover change and was better than the 1990 classified image. In this study, the manufacturer and user accuracy ranged from 82.35 to 93.75% and 85.29 to 93.75% in the year 2000, from 81.25 to 90.63% and 80.95 to 91.43% in the year 2010 while in 2020, it ranged from 88.57 to 96.77% and from 89.74 to 96.88%, respectively. The kappa coefficient of classification was 85.93, 85.52, and 91.59% in 2000, 2010, and 2020, respectively. Thus, according to the classification scale given by Moriasi et al. (2007), the classification is on the very good grade and excellent scale in the years 2000, 2010, and 2020, respectively.

9.4.2 Land Use and Land Cover Maps

Spatial analysis was performed to describe land cover change patterns and overall land use changes over time. Water bodies, built-up areas, forestlands, shrublands, grasslands, companion lands, and barren lands were the major land use and land cover classes of the study area. Land use–land cover change of Panchet dam area (Figs. 9.2 and 9.3) converts most of the barren and felled land into shrubland, residential land with forest areas. This result shows that the present zone was shifted from a barren zone to a green zone.

Analysis of the results indicated that fellow lands were the predominant land cover class with 28.27 and 33.20% in 1990 and 2000 whereas barren land also covers most of this area with 13.94 and 30.21%, respectively. An analysis of the 2010 and 2020 images showed that the proportion of grassland was the largest in 2010 and decreased by 15.45% in 2020, while shrubland had the second major coverage of 20.46% after grassland in 2010. In the year 2020, shrubland had a major coverage of 33.57%, and this year the forest area had the second major coverage. This result is due to changing phenomena and the government's action toward natural resource conservation and management. In general, the results indicate that forestlands, shrublands, and built-up areas have increased while grasslands, companion lands, water bodies, and wastelands have decreased.

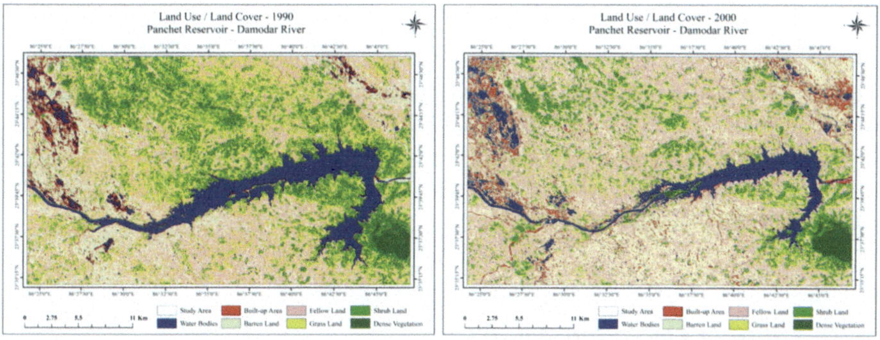

Fig. 9.2 Land use–land cover map of 1990 and 2000 of the study area

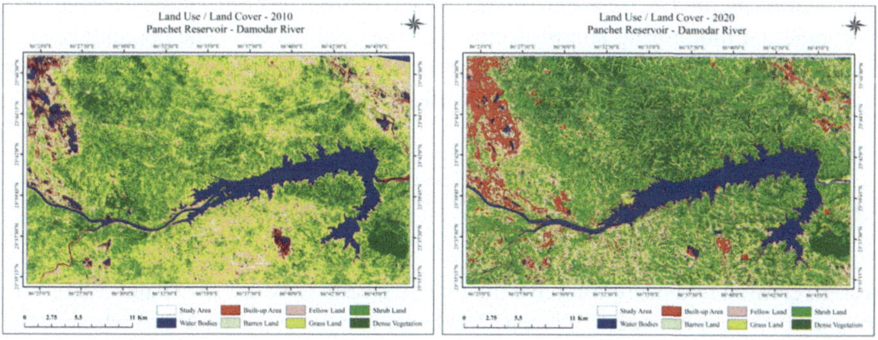

Fig. 9.3 Land use–land cover map of 2010 and 2020 of the Study area

Table 9.2 Summary of land cover classes in the Panchet Dam region from 1990 to 2020

Year	1990		2000		2010		2020	
LULC class	Area (km^2)	%	Area (km^2)	%	Area (km^2)	%	Area (km^2)	%
Water bodies	8832.96	9.86	6640.47	7.41	7864.65	8.77	7760.05	8.66
Built-up area	2880.45	3.21	5077.98	5.67	2685.87	3.00	6955.23	7.76
Barren land	12,490.11	13.94	27,077.40	30.21	4490.28	5.01	7211.49	8.05
Fellow land	25,335.00	28.27	29,754.45	33.20	9368.64	10.45	9630.51	10.74
Grass land	21,995.10	24.54	9931.05	11.08	36,616.68	40.85	13,851.67	15.45
Shrub land	12,383.28	13.82	7665.12	8.55	18,336.06	20.46	30,086.36	33.57
Dense forest	5712.30	6.37	3482.73	3.89	10,267.02	11.45	14,133.63	15.77

LULC changes in the region revealed that forest and shrubland area increased over the past 30 years from 1990 to 2020 (Table 9.2) while other classes decreased. The water bodies were creating uniform coverage. This may be due to the formation of large reservoirs in the basin. This land use and land cover change can affect the hydrology of the catchment area.

9.4.3 *Change Detection Analysis*

Detection of land use and land cover change based on remote sensing images has been widely applied in research for LULCC, natural resource management, and environmental monitoring and protection (Zhang et al., 2009). The percentage area of each land cover class was obtained for each year separately from supervised classified images using Erdas Imagine software integrated with ArcGIS (Fig. 9.4).

9.4.4 *Erosion and Accretion*

Overall, erosion and accretion have fluctuated greatly between 2000 and 2020. The amount of erosion between 1990 and 2000 was very high (3299.20 ha) whereas the amount of accretion is only 224.09 ha. If we talk about the no-changing area, about 4078.65 ha area remains as it is, not much has changed (Fig. 9.5).

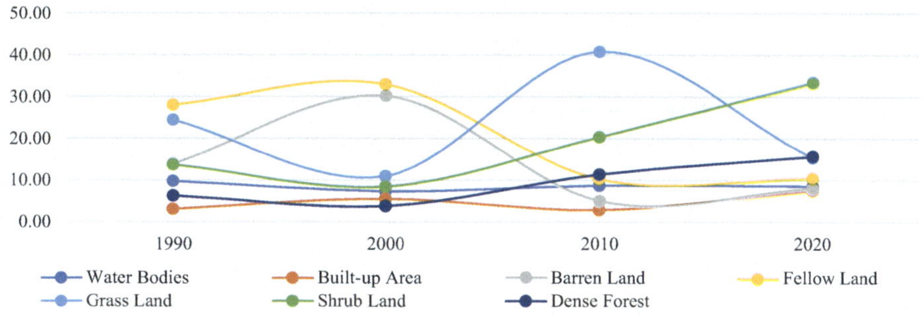

Fig. 9.4 Land use–land cover area% of Panchet Dam region during 1990–2020

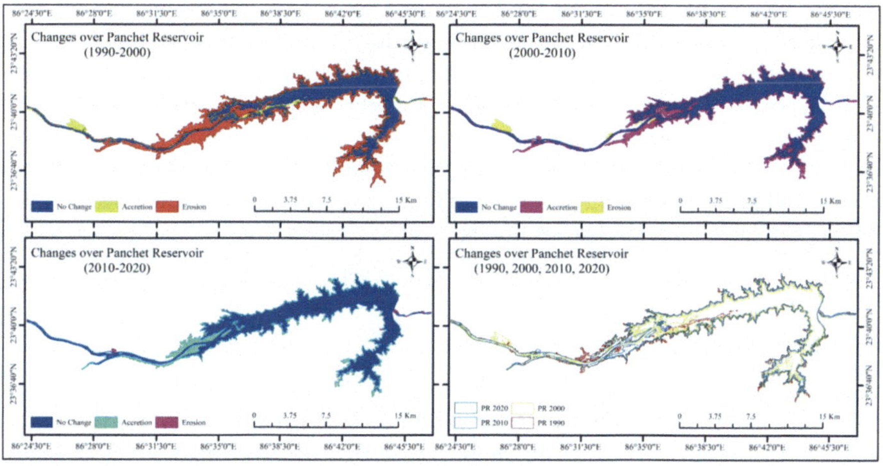

Fig. 9.5 Erosion and accretion of Panchet Dam from 1990 to 2020

In contrast, from 2000 to 2010, the amount of accretion was comparatively higher than erosion, where erosion was only 182.69 ha and accretion was 1761.88 ha. The no-changing area was affected by the time zone of the year 2000 and covered 4120.05 ha. The accretion was observed to be very high between 2010 and 2020, while the extent of erosion was also found to decrease as compared to the previous year, which is a good sign. Talking about the no-changing area, there is a lot of about 5809.56 ha in this decade (Table 9.3 and Figures 9.5, 9.6).

Table 9.3 Erosion and accretion area (Ha) of Panchet Dam from 1990 to 2020

Years	Erosion	Accretion	No-change area (Ha)
1990–2000	3299.20	224.09	4078.65
2000–2010	182.69	1761.88	4120.05
2010–2020	72.37	1118.47	5809.56

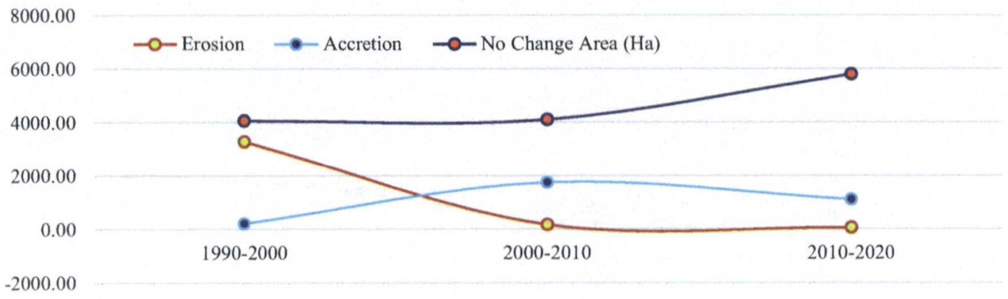

Fig. 9.6 Erosion and accretion area of Panchet Dam from 1990 to 2020 - Graph

The changes observed during the temporal analysis of the Panchet Dam has a significance over the local sustainability of the region. The decrease in the erosional area emphasizes on the initiatives taken by the local communities and authorities in preserving the local vegetation and biodiversity near the dam. However, the increase in the deposits of several layers of matter brought by the river has increased; i.e., accretional process has increased temporally. The Damodar River is carrying and depositing the amount in the form of pebbles, clay, stones, etc. This accretional process will going to be a big challenge in future as it will interrupt the flow of water and this can lead to the disastrous floods.

9.5 Conclusion

This study has analyzed the LULC detection of Panchet Dam over the past three decades. The results of the study showed that significant changes were detected during the study period. Forest and shrubland have increased in this decade, while other classes showed a decreasing trend. The result indicates that the conversion of the class areas can lead to problems including changes in river flow, hydrological systems, and soil degradation in the region. These changes can have an impact on sustainable resource management and the livelihoods of local societies. Therefore, better land management practices (soil and water conservation), better agricultural inputs, integrated watershed management (land use planning and management), and active participation of the local community should be advanced to prevent unwanted LULC dynamics in the basin. In this study, change detection analysis using GIS and remote sensing can provide useful information for planners and decision-makers to understand the seasonal patterns of land use dynamics resulting in sustainable land management planning.

Forest degradation has become a major issue in developing countries in recent decades. In 2000, the total area of degraded forest in 77 countries was 800 million hectares (ha), of which 500 million hectares had changed from primary to secondary vegetation. The reduction of forest cover at community limits in developing world contexts, such as in this study, would have implications in both global biodiversity and global warming contexts. This study has provided indicative data on a local (large) spatial scale. Such data may provide a basis for monitoring changes in tree cover over the long term, which is important to subsequently provide threshold-indicated alarms based on established conversion rates.

References

Anderson, J. R. (1976). *A land use and land cover classification system for use with remote sensor data* (Vol. 964). US Government Printing Office.

Bajocco, S., De Angelis, A., Perini, L., Ferrara, A., & Salvati, L. (2012). The impact of land use/ land cover changes on land degradation dynamics: A Mediterranean case study. *Environmental Management, 49*(5), 980–989.

Baynard, C. W. (2013). *Remote sensing applications: Beyond land-use and land-cover change.*

Bid, S. (2016). Change detection of vegetation cover by NDVI technique on catchment area of the Panchet Hill Dam, India. *International Journal of Regularization Geography, 2*(3), 11–20.

Bid, S., & Siddique, G. (2019). Identification of seasonal variation of water turbidity using NDTI method in Panchet Hill Dam, India. *Modeling Earth Systems and Environment, 5*(4), 1179–1200.

Bid, S., & Siddique, G. (2020). Water-level fluctuation (WLF) of Panchet dam in India and assessment of its human risk using AHP method. *GeoJournal, 12*, 1–26.

Bird, E. C. F. (1983). Factors influencing beach erosion and accretion: A global review. *Sandy Beaches as Ecosystems, 321*, 709–717.

Dewan, A. M., & Yamaguchi, Y. (2009). Land use and land cover change in Greater Dhaka, Bangladesh: Using remote sensing to promote sustainable urbanization. *Applied Geography, 29*(3), 390–401.

Foody, G. M. (2002). Status of land cover classification accuracy assessment. *Remote Sensing of Environment, 80*(1), 185–201.

Ishtiaque, A., Shrestha, M., & Chhetri, N. (2017). Rapid urban growth in the Kathmandu Valley, Nepal: Monitoring land use land cover dynamics of a himalayan city with landsat imageries. *Environments, 4*(4), 72.

Kibret, K. S., Marohn, C., & Cadisch, G. (2016). Assessment of land use and land cover change in South Central Ethiopia during four decades based on integrated analysis of multi-temporal images and geospatial vector data. *Remote Sensing Applications: Society and Environment, 3*, 1–19.

Kumar, R., & Saikia, P. (2020). Forest resources of Jharkhand, Eastern India: Socio-economic and bio-ecological perspectives. In *Socio-economic and eco-biological dimensions in resource use and conservation* (pp. 61–101). Springer.

Lu, D., Mausel, P., Brondizio, E., & Moran, E. (2004). Change detection techniques. *International Journal of Remote Sensing, 25*(12), 2365–2401.

Lu, D., & Weng, Q. (2007). A survey of image classification methods and techniques for improving classification performance. *International Journal of Remote Sensing, 28*(5), 823–870.

Ma, L., Li, M., Ma, X., Cheng, L., Du, P., & Liu, Y. (2017). A review of supervised object-based land-cover image classification. *ISPRS Journal of Photogrammetry and Remote Sensing, 130*, 277–293.

Moriasi, D. N., Arnold, J. G., Van Liew, M. W., Bingner, R. L., Harmel, R. D., & Veith, T. L. (2007). Model evaluation guidelines for systematic quantification of accuracy in watershed simulations. *Transactions of the ASABE, 50*(3), 885–900.

Parker, D. C., Manson, S. M., Janssen, M. A., Hoffmann, M. J., & Deadman, P. (2003). Multi-agent systems for the simulation of land-use and land-cover change: A review. *Annals of the Association of American Geographers, 93*(2), 314–337.

Rashid, I., & Aneaus, S. (2020). Landscape transformation of an urban wetland in Kashmir Himalaya, India using high-resolution remote sensing data, geospatial modeling, and ground observations over the last 5 decades (1965–2018). *Environmental Monitoring and Assessment, 192*(10), 1–14.

Short, N. M. (1982). *The Landsat tutorial workbook: Basics of satellite remote sensing* (Vol. 1078). National Aeronautics and Space Administration, Scientific and Technical Information Branch.

Shukla, S., & Gedam, S. (2019). Evaluating hydrological responses to urbanization in a tropical river basin: A water resources management perspective. *Natural Resources Research, 28*(2), 327–347.

Tewabe, D., & Fentahun, T. (2020). Assessing land use and land cover change detection using remote sensing in the Lake Tana Basin, Northwest Ethiopia. *Cogent Environmental Science, 6*(1), 1778998.

Yesuph, A. Y., & Dagnew, A. B. (2019). Land use/cover spatiotemporal dynamics, driving forces and implications at the Beshillo catchment of the Blue Nile Basin, North Eastern Highlands of Ethiopia. *Environmental Systems Research, 8*(1), 1–30.

Zhang, X., Hu, Y., Zhuang, D., Qi, Y., & Ma, X. (2009). NDVI spatial pattern and its differentiation on the Mongolian Plateau. *Journal of Geographical Sciences, 19*(4), 403–415.

Chapter 10
Tourist's Satisfaction and Sustainable Cities: Case Study of Udaipur City, India

Sabiha Khan, Prem, and Umesh Kumar Jeengar

Abstract The study focuses on measuring the satisfaction levels of tourists visiting Udaipur city, as tourism is a crucial factor for the sustainability of the city. The research is to assess the level of enjoyment that visitors experience and to determine the crucial components that affect both the sustainability of the region's tourism business and the overall visitor experience. To this end, 278 respondents were interviewed based on ten satisfaction criteria. Along with the city's natural and cultural attractions, the study examines several aspects of Udaipur's tourism infrastructure, such as accommodation, standard amenities, security, regional cuisine, and transit options. Structured interviews have been conducted at the selected tourist spots of the city, based on the Department of Tourism, Rajasthan. The data gathered was then statistically evaluated to classify replies into three levels: highly satisfied, satisfied, and dissatisfied. The results show that visitors were rather pleased with Udaipur's abundant scenic surroundings, rich cultural heritage, and quietude atmosphere, accessibility to lodging, and basic facilities with secure environment. Meanwhile, there was significant discontent with the quality of regional cuisines, the reliability and accessibility of tour and guide services, and the city's transportation infrastructure and traffic management. This study helps to work for the areas of concern for the sustainable tourism in the city. This study helps to work for the areas of concern for the sustainable tourism in the city.

Keywords Tourism satisfaction · Sustainable · City problems

S. Khan (✉) · Prem · U. K. Jeengar
Department of Geography, Faculty of Earth Sciences, Mohanlal Sukhadia University, Udaipur, India
e-mail: drsabihakhan@mlsu.ac.in

© The Author(s), under exclusive license to Springer Nature Switzerland AG 2024
B. V. Raj Singh and A. K. Batar (eds.), *Sustainable Local Development for Environmental and Social Sustainability*, Human-Environment Interactions 11,
https://doi.org/10.1007/978-3-031-67303-0_10

10.1 Introduction

Sustainability is a societal goal that relates to the ability of people to safely coexist on earth over a long time. Among the 17 SDG goals comprise with three dimensions: social, environmental and economic, eleventh goal is related with Sustainable Cities and Community to make cities and human settlements inclusive, safe, resilient and sustainable. It needs new, innovative, smart and intelligent urban planning as the population of the world is constantly increasing, especially urban areas (United Nations, 2015). Cities are dreamland for rural inhabitants; people move to urban areas for the betterment of their lives and to avail better opportunities of job, health and education facilities. This over migration in urban areas leads to cities overcrowded with population and getting crammed beyond its capacity day by day. In this situation, people tend to compete over the limited and scarce resources such as living space, electricity, water, transport, health and education facilities; on other hand, multiple health risks and socioeconomic problems were major issues of concern (UNEP, 2019). For the sustainable development of any city, it is necessary to encourage the base industry there, because it is the most important component in the economic development of the city and it is also necessary to eradicate the hindering factors present in the way of development. In India, the Smart City Mission plan related to this was already started in 2015 itself. Ministry of Home Affairs and Urban Affairs selected 100 cities of the country for this project, and Udaipur city is one of them. It is an urban renewable and retrofitting program to develop smart cities and make them sustainable for the citizens. (https://smartcities.gov.in/) Tourism, handicraft and mineral industries are important in the economic stability and development of Udaipur city. But tourism, along with being the main industry of the city, is also the main identity of Udaipur. It creates so many opportunities of direct and indirect employment for most of local people and is the basic contributor of the city's economy. Nowadays, the city is hot spot of pre-wedding shootings and destinations weddings. Today, tourism industry of the city is a thousand crore industry with more than a thousand of hotels, restaurants, resorts, homestays and guest houses.

10.1.1 Factors Affecting Tourism in Udaipur City

There are so many factors that determine tourism and influence the running of the tourism industry. That particular factor turns the place as a tourist destination. Some of them show immediate effect while other factors affect in the long run (Pizam & Neumann, 1978). Among them, there are many such factors which make the city of Udaipur a perfect tourist destination:

A. **Environment:** Climate and weather season act as a great magnet to attract tourists and enable them to enjoy the gift of nature can be encashed to promote tourism. Monsoons in Udaipur are truly magical and breathtaking owing to the earthy freshness. It is a particularly lovely time to bask in the beauty of this wonderful natural phenomena and lush green landscapes. Winter season is also regarded as pleasant and comfortable for sightseeing and exploration of the natural, cultural and heritage sights.

B. **Landscape Resources**: Physiography and geology are major components of landscape resources, and it provides different kinds of related activities. Udaipur has some related activities like kayaking, boat riding, trekking, rope way and zip inning adventures and jungle safari and so on that are maintaining the enthusiasm in the tourists.

C. **Historical, Cultural and Religious Importance**: Rich historical and cultural heritages have great significance to attract tourists to the destination. Udaipur has a significant asset of heritage in its historic core with tangible and intangible forms. The city's heritage boosts not only palaces, ghats, temples, havelis but also for other attractions such as the historic core which is a living heritage where markets and residential clusters coexist. The places of religious importance are also act as tourist spot, where tourism is at its peak at particular time period in a year. Jadish Temple, Mahakaleshwar Temple, Neemuch Mata, Karni Mata, Mastaan Shah Dargah are places of religious interest. Archeological sites, specific physical features, international sports and cultural events are other factors of tourist attraction.

10.1.2 Positive and Negative Impacts of Tourism

Tourism is one of the fastest growing sectors in the world, which spread its positive and negative impacts on the economy, infrastructure, and culture–social and natural environment of a nation (Mdusman, 2016).

A. **Economic Implications:** Tourism is one of the most flourished industries in the world, and it generates a lot of revenue every year through direct taxation and indirect taxes on goods and services availed by the tourists. This sector provides numerous employment opportunities to the local people and one of the top five job creation sectors in the world (Pal & Rawal, 2020). However, many jobs in the tourism industry have been criticized as they usually poorly paid and many people engaged in this industry find themselves unemployed during the off peak seasons (Rahman, 2017).

B. **Infrastructure:** Through private and public initiatives, a destination may see a lot of infrastructural development to cater for the needs of the visitors. To boost huge tourism, money is invested in preserving the local heritage, improving infrastructure and providing better local facilities. Overdevelopment comes at the cost of nature. However, mass tourism is often a cause of overcrowding and traffic jams that disrupt the lives of local people. Many attractions and the infrastructure of the destination may also suffer wear and tear due to heavy tourist presence and movements (Rahman, 2017).

C. **Society and Culture:** Tourists visit different places to explore new culture of related regions, and sometimes exchange of culture creates cosmopolitan culture in the region. Their mutual interactions offer opportunities for better understanding and removing any misconceptions and prejudices. However, there is also a risk that both tourists and locals may hurt each other feelings due to a lack of understanding and sometimes may trouble the local culture and create unrest (Rahman, 2017).

D. **Natural Environment:** A host destination can spend money generated from tourism on protecting natural attractions. National parks, bioreserves, sanctuaries are such beautiful places, where flora and fauna are preserved without any interference. According to the BBC Report (2019), tourists are often held responsible for causing air, water and noise pollution, topographic erosion, littering and other natural environmental damage to a tourist destination. Due to the excessive pressure of tourist football on these areas, topographic features and diversity tend to lose its original form. Tourist activities like campfire, forest safari, trekking, water sports and so on carried out a negative effect on the natural components, degrade the environment by pollution and disturb the ecological balance of the region (Rahman, 2017).

Natural beauty and cultural heritage are unique features in a beautiful tourist city like Udaipur. There are some positive and negative factors in the development of every city; naturally, they are also present here in this city. Tourist satisfaction is the mainstay of the industry in the tourist city. Therefore, an attempt has been made in this paper to assess the satisfaction of tourists about the city of Udaipur.

Various researches have been conducted to find out the root factor behind the satisfaction of tourist at particular destination. A study was attempted at Dubrovnik-Neretva County to the find the appropriate model for the tourist satisfaction to maintain the competitiveness in the global market, and significant correlation has been observed between four independent variables (Pavlic, 2015). There were so many studies, conducted on tourism in Thailand; one on them was based on service provider industry for tourism and another study revealed that cost of staying at Chaing Mei was the most attractive factor for the tourist's attraction (Suanmali, 2014). In a study, 13 items of motivation and eight items related to the activities were used at Andalusia, Spain (Armario, 2008). And another study was based on eight attributes factors and 30 subattributes to examine the relationship between overall tourist satisfaction and destination attributes at Klang Valley, Penang and Malacca, Malaysia (Shahrivar, 2012). The research was conducted on the tourist's visited Ho chi Minh city to find out the destination satisfaction and tourist's return intention; on the basis of independent variables, it concluded that the majority of foreign tourists relaxed and enjoyed their leisure activities with their family and friends (Khuong & Nguyen, 2017). A structural equation modeling approach has been applied on the tourists of Kangyang tourist spot to find tourist satisfaction and willingness to revisit accommodation including with the concepts of ecological resources, health preservation and tourism activities (Liyun Zeng, 2021). A research was conducted to find out the attraction and ecotourism and toward the willingness of revisit, and the significant result between the above variables has been found in Sumatra (Trinanda & Sari, 2020). Another study emphasized on the assessment of tourist's satisfaction factors at Hamadan city, and it has been observed that society behavior, municipal equipment and cost of services were the top three parameters (Leili Tapak, 2019) whereas social security was found to be the most important factor of tourist satisfaction at Langkawi Island, Malaysia, and followed by tourist expectations, destination image and cost and risks (Aliman, 2016). On Rajasthan, Respect, Trustworthiness Dedication and Care (RTDC) model with nine components has been suggested o Rajasthan Tourism Development Corporation Ltd., Government of Rajasthan, for the attraction and satisfaction of tourist (Jingar, 2015). A study examined the tourist's overall satisfaction and the return intention for the destination on the basis of the tourist's satisfaction and dissatisfaction at the Majorca island, part of the Baleric Islands (Alegre & Garau, 2010). Structural equation modeling has been used to identify the variables influencing visitors' happiness in the city of Nha Trang and to assess the level of satisfaction and loyalty among the European tourists had for their travel destinations, and the study revealed that friendliness and hospitality of local people were the main satisfactory factors of tourists (Le & Dong, 2017). Multiple regression models were used to examine the key determinants of overall satisfaction and destination loyalty at nature-oriented tourism destination in the Flinders Ranges region, South Australia. Study revealed that tourists seek to discover new and adventurous experiences for satisfaction. It has also been also observed that the landscape or scenery attribute was found to be a very important motivational as well as satisfaction factor for tourists (Kim & Brown, 2012). Revisit intention of tourists has been studied in Hong Kong, based on their evaluations of destination performance more than levels of satisfaction. (Um et al., 2006) Structural equation modeling and a multi-group analysis were used to examine the effect of gender and visit frequency as a moderator variable on the relationship among the destination's perceived quality, tourist's satisfaction, and word-of-mouth in Damang City (Wang et al., 2017).

10.2 Study Area

Udaipur is well known destination on the global tourist map of the world and better known for its breathtaking landscapes, history, culture, festivals and heritage structures. Udaipur is ringed by gentle Aravali hills and having blue dotted beautiful lakes in it. This scenic beauty makes it, "the Venice of east" and "Kashmir of Rajasthan". Rainy, post-rainy and winter seasons are very pleasant and most appealing for visit (Fig. 10.1).

The main contributors to the economy of the city are tourism and agriculture sectors and mineral industries. The cottage and handicraft industries play an important role toward the economic growth of the city. The city also has been selected for the smart city development mission by the Government of India's initiative. Tourism sector has been not only the major contributor to the economic growth of the city but it also popularized the city in tourism industry of the world (Udaipur times).

Cultural heritage of the city has balance between rituals and traditions of the past and modern lifestyles and advancements. On one side, vibrant folk dances like

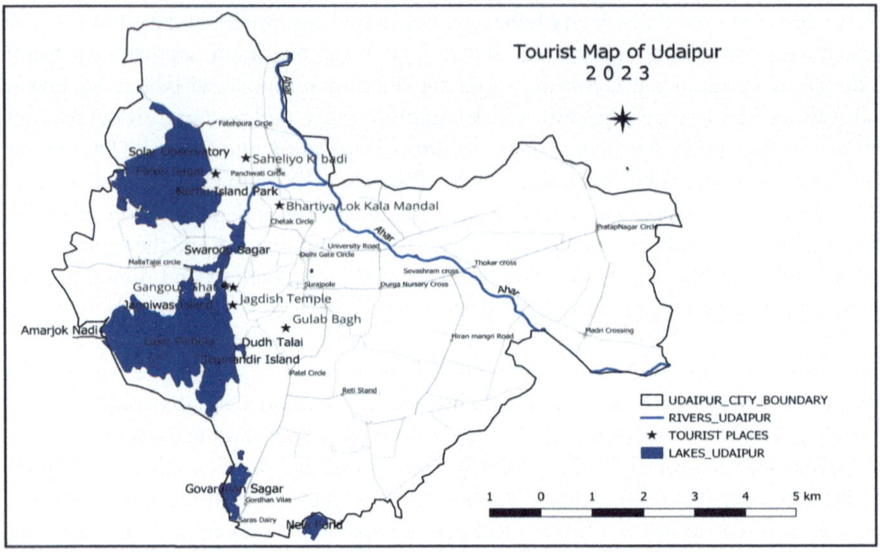

Fig. 10.1 Study area map of Udaipur city. *Source* Govt. of Rajasthan, India

Ghoomar, Kalbelia, Kachchhi Ghodi, folk music with traditional instruments and miniature art, gangaur festival, shilpgram utsav, etc. add a sparkle of richness to the cultural heritage of the city. On the other hand, world music festivals and Udaipur lake festival add to this marvelous and unmarked hospitability which turns out it one of the most expensive destinations in the world for weddings, which add a feather in the cap. Udaipur is also the host city for the first ASEAN Art camp. Recently, the first G20 Sherpa Meeting was also hosted by Udaipur city (Times of India, 2022). Udaipur is well connected to the major cities of India by land, rail and air. The city lies on the intersection point of East–West Corridor, Golden Quadrilateral, National Highway (NH) 76 and National Highway (NH) 8, midway between Delhi and Mumbai, located around 700 km from either city. Air connectivity is increasing nowadays; it also contributes in flow of tourist football in the city.

According the released data by the tourism department, the domestic tourist has broken all records in year 2022. This is the first time in the history when city has not only observed such huge football of domestic tourist; it also witnessed a significant growth in number of foreign tourists. The constant growth in the tourist football is a healthy sign for tourism industry of the city after recent tensed past of corona pandemic, and it is also expecting increase in numbers of tourists during the forthcoming years (Fig. 10.2).

There is a vast scope for promoting rural, ecotourism and agro-tourism around the city. Udaipur's identity as a tourist hot spot has grown multifold on the global map in the recent days (cntravels.com).

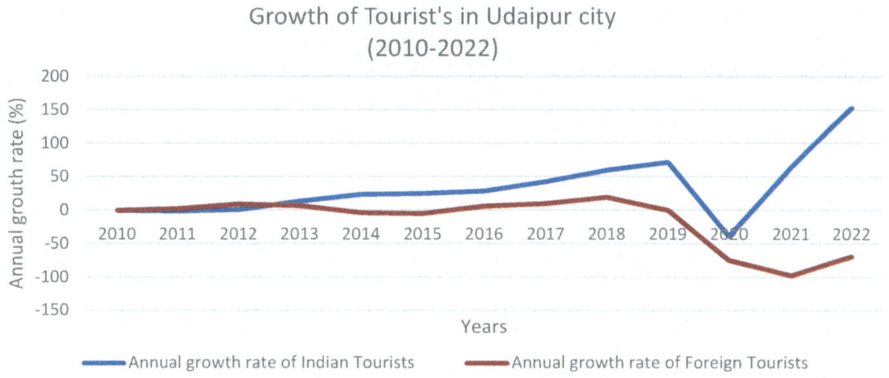

Fig. 10.2 Growth of tourist's in Udaipur city. *Source* Tourism Department, Udaipur

10.2.1 Methodology

This study has adopted quantitative approach in data collection and analysis. Major tourist spots have been selected for data collection, which has been provided by the Tourism Department, Government of Rajasthan (Fig. 10.3).

To measure the tourist destination satisfaction of tourists, quantitative approach has been used. For this, 278 respondents were interviewed on the basis of ten criteria of satisfaction: local city attraction (natural and cultural heritage); accommodation and basic amenities (water, electricity, washroom, internet, etc.); safety and security; entertainment and recreational activities; local market; local cuisines; tour and travel guides; quietude and peace; spot accessibility and transport facilities; and road infrastructure and traffic management (Fig. 10.4).

Responses for tourist destination satisfaction from the tourists based on the above criteria have been classified in three categories, viz. very satisfied, satisfied and dissatisfied. Results have been presented in the form of statements with the statically values obtained from their replies of particular criteria.

10.3 Result and Discussion

Satisfaction is the focal point of any tourist's behavior which is helpful in determining the destination choice by the individual. Satisfaction is aftereffect of the relationship between expectation of tourist about the destination and result of their involvement with that destination point. Results of above-mentioned ten criteria of tourist destination satisfaction have been presented below:

Accommodation and basic amenities (water, electricity, washrooms, internet, etc.): 90% of the tourists were found to be very satisfied with the hotel and basic facilities, while 8 and 2% were found to be satisfied and dissatisfied, respectively.

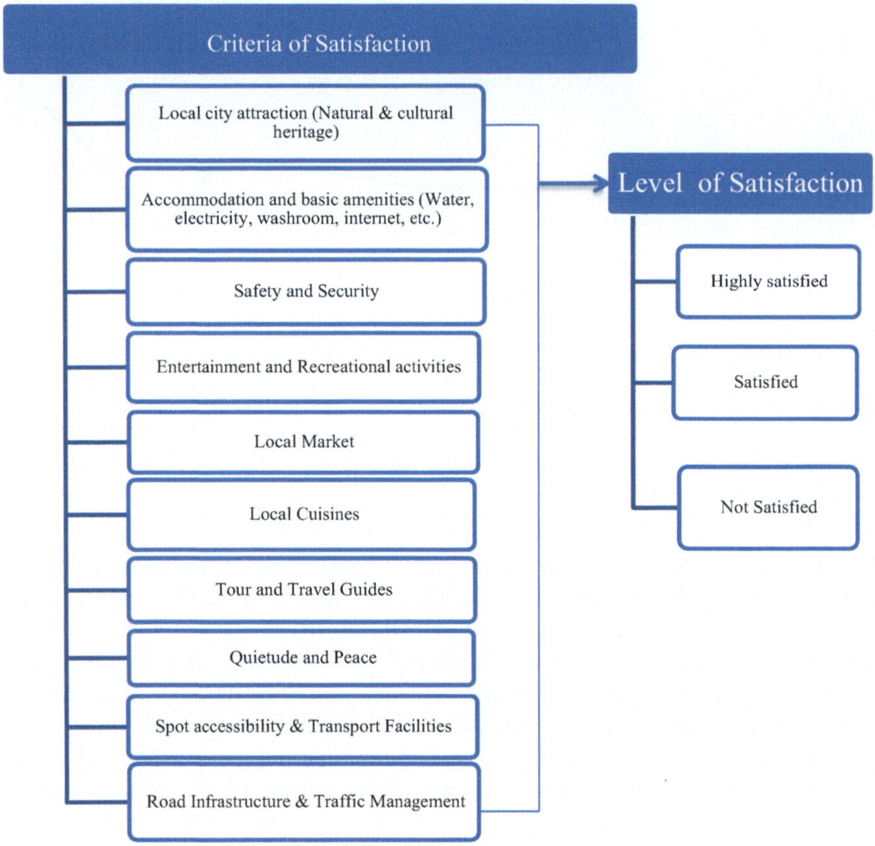

Fig. 10.3 Methodology flowchart

Along with various five star and seven star hotels, there are many other good hotels available which are fully equipped with all facilities. So most of the tourists, especially people belonging to high-income groups, are completely satisfied, while these types of hotels are beyond the reach of the middle-income group, and their budget hotels are not equipped with such facilities, so they are not satisfied. Late night arrival of tourists or tourists with no pre-booking are charged heavily irrespective of their quality of facilities provided which is the main reason of dissatisfaction of tourists.

When it comes to safety and security, 88% of the tourists came here are very satisfied. This city was found to be more secure than other cities. Most of the tourists who were very satisfied are foreigners. 9% of the tourists who feel safe in roaming the city are people who are staying in the hotels near the tourist places as these places are having activities throughout the day. But on the other hand, the tourists who are staying in peripheral hotels or homestays or who are aware of past misdeeds in the city were feeling insecure, and they are 3% in total.

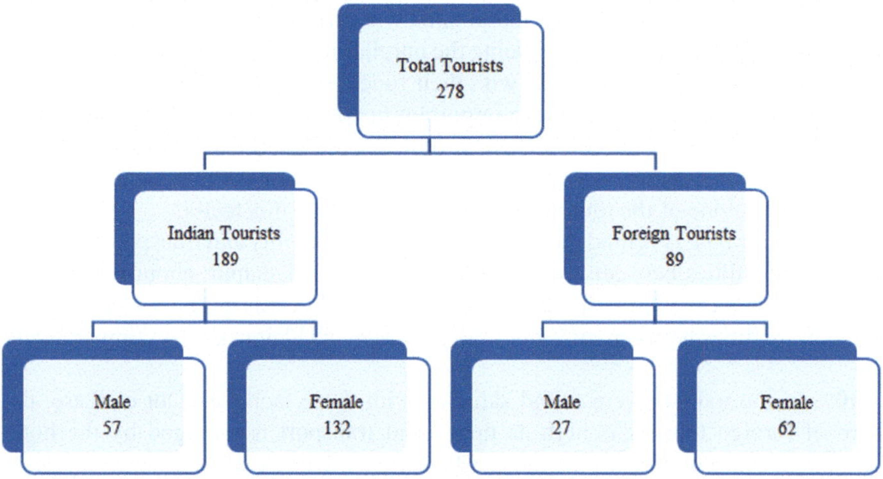

Fig. 10.4 Tourist sample data

The survey conducted in this context on entertainment and recreational activities study revealed that 42% of the tourists did not have knowledge about other recreational activities other than boating that are available in the city. Remaining 58% of the tourists and out of these 83percent tourists who availed these activities were found satisfied and 17% were found unsatisfied with the pollution in lakes and due to the irregularities in activities at other tourist places.

The percentages of highly satisfied tourists with the local market were observed 28%, out of which most of the tourists were foreigners. The main reason for this is the availability of various types of handicraft items available in the city, such as leather, wooden, junk jewelry and marble items. 60% of the tourists were found satisfied with the varieties available in the city, while 12% tourist are unsatisfied, mainly due to price difference of the same product within the city and between the cities.

Only 14% tourists were found satisfied in terms of local cuisines, most of whom were foreign tourists and whose food was available only in five star or seven star hotels. Domestic tourists from other states who tried local cuisines here for the first time were found satisfied. 86% tourists were found unsatisfied with the local cuisine, the main reason being issue with food hygiene, food quality and service quality.

According to the survey conducted in the context of the city's natural beauty and historical heritage, 89% tourists were found to be very satisfied with the tourism importance of the city, out of which domestic tourists were more enthusiastic about natural beauty and foreign tourists were found more enthusiastic about historical heritage, while 8% of tourists were found satisfied. Only 3% of tourists were dissatisfied due to incomplete knowledge of features available at historical tourist places and neglect of cleanliness at natural places.

28% of the tourists were found dissatisfied with the importance of tour and travel guides in tourism, the main reason being the unreliability of their work, whereas 18% of the tourists were found satisfied with their functioning.

Quietude and peace are also very important factors for tourist to relax their mind and bodies. Along with the natural beauty of the city, 86% of the tourists were found to be very satisfied and 14% were found to be satisfied with the experience of peace here, whereas none of the tourists were dissatisfied with this factor.

There are two facts behind the concept of spot accessibility and transport facilities: One is the facilities between the tourists' hometown to Udaipur, about which most of the tourists were found satisfied, whereas the level of transport accessibility and transport facilities between different tourists places in Udaipur city is comparatively low.

40% of the tourists were found satisfied with these facilities. Out of these, the share of foreign tourists is high as their local transport is arranged by the hotel administration, which is well organized. Easy availability of bike and bicycle at tourist places is another major reason for satisfaction whereas online booking for cab services is not adequate here and charges of local transportation are also very much high, that is why 58% of tourists were not satisfied with this facility and only 2% were found highly satisfied.

Only 4% of the tourists were found satisfied with the road infrastructure and traffic management system in the city while the other 96% were found dissatisfied with the system. Haphazard road side parking, road side encroachments, inadequate parking spaces, vulnerable road users with abiding of traffic rules and regulations, lack of street lights, traffic signals and pedestrian tracts are reasonable factors behind the dissatisfaction.

Data represent that maximum respondents were satisfied with the all the criteria of tourism available in the city. There were some aspects in which tourists were found highly satisfied like the importance (natural landscape and cultural heritage) and quietude and peace of the city. They were also very much satisfied with available accommodation and basic facilities and found city safe and secure as compared to other tourist cities.

But it has been also observed that tourists found unsatisfied about available quality of local cuisines and tour and guide facilities. They found inadequate transport facilities and traffic management system, which is the matter of concern for the sustainability of the tourism in the city (Fig. 10.5).

The satisfaction level of tourists is very important in the operation of any tourism industry. There are so many initiatives completed by Udaipur Smart City Limited, and their direct and indirect effects can be observed on sustainable tourism in the city.

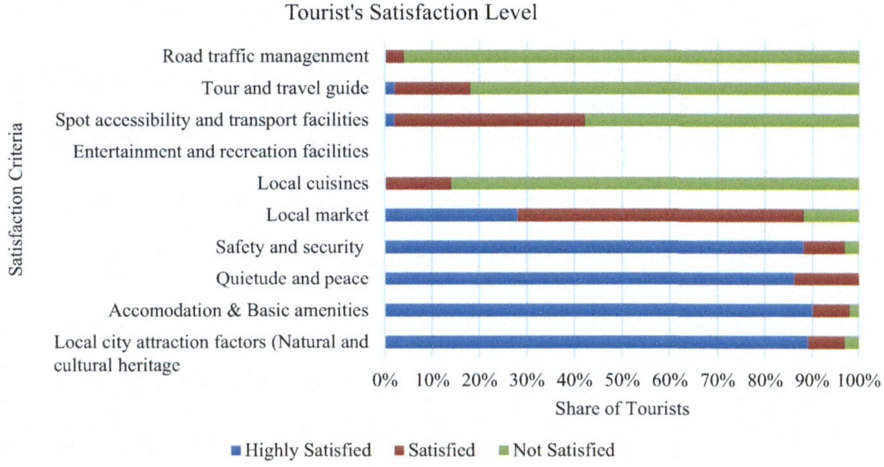

Fig. 10.5 Results of tourist's satisfaction level. *Source* Compiled by Researcher

Sr. No	Smart cities mission—objectives	Efforts of Udaipur smart cities limited	Motivation to tourism development and sustainability
1	Adequate water supply	Water ATM, water quality monitoring system	Basic need of tourism and cleanliness
2	Improved sanitation	Progress under SBM	Need of a tourist and nudge to keep the place clean
3	Solid waste management	Bioremediation at Balicha Collection Vehicle commissioned under SBM	Need of the area and nudge to tourists to keep the place clean
4	Efficient urban mobility and public transport	Operation of e-rickshaw, parking lot development nearby at PWD garage near Gulab Bagh, microlevel intervention in walled city areas, junction improvement of Surajpole, construction of bridge near Daiji ki Pulia, multilevel parking at Chandpole, construction of parking lot at RNT and MB Hospital	Reduces carbon footprint in tourism

(continued)

(continued)

Sr. No	Smart cities mission—objectives	Efforts of Udaipur smart cities limited	Motivation to tourism development and sustainability
5	IT connectivity	Construction of command and control building, installation of around 500 Wi-Fi hot spots	Promotes better connectivity to ensure the marketing, accessibility and research activities in tourism
6	Sustainable environment	Pollution control in Pichola Lake by trapping wastewater coming from Sisaram Village, water quality monitoring system, construction of Bird Park through convergence	To motivate to contribute to the achievement of Sustainable Development Goals, especially Goal 6
7	Safety and security of the residents	Construction of command and control building	Attract more tourists and to achieve Goal 16 in regard to keeping peace and strength
8	To improve the quality of tourism destinations	Conservation of step wall, renovation of Ghats and waterfronts, Ayad River Front Development, Design, Develop and Manage adventure zone at Manikyalal Verma Udhyan Park	The attraction for the tourists

Source of data for smart cities mission—initiatives: official website of ministry of housing and urban affairs smart cities mission statement and guidelines
Source of data for getting information about efforts of Udaipur smart cities limited: official website of Udaipur smart city limited

फतहसागर को सींचने वाली नहर में जल से पहले बहा पसीना

Source: Rajasthan Patrika, 08, May' 2023

Several commendable works have been done by the government, non-governmental organizations and local people in maintaining the sustainability of city's base industry—tourism, which are as follows: Under the National Lake Conservation Project, along with the rescue of Lake Pichola and Fatehsagar, the work of bio-conservation and beautification of the surrounding area has also been done. Jheel Sanrakshan Samiti (JSS), an Udaipur based non-government organization, is working for the conservation of water bodies since 1982. Along with this, regular contributions are being made by the local people for the cleanliness and protection of the lakes.

Source: www.india.mongabay.com

Rajasthan Tourism Department has taken several steps under Rajasthan Tourism Policy-2020 to reduce the pressure on local tourist places to create awareness about ecotourism destinations and lesser known destinations. Under this, ecotourism projects like Sajjangarh Biological Park, Badi Jheel and Purohito Ka Talab have been prepared and successfully operated, along with a plan for their expansion that has also been proposed. In a community-led project run by the Urban Forestry Scheme (Nagar Van Udyan), 31 youths from the local community have been given basic training and made in-charge of the ecotourist site. Under this, a "Valley of Flowers" has been created on 3.7 km^2 of unusable and infertile land of Chirwa village. An ecotourist circuit has been built near it, including Mewar Biodiversity Park and Purohito Ka Talab. Today, this project is not only becoming economically beneficial, but has also proved effective in conservation and development of forest resources.

10.4 Conclusion

Udaipur city holds an important place in the world in terms of tourism, and this sector is the major contributing factor in the economic development of the city. Udaipur city has mesmerizing natural beauty and royal historical and cultural heritage. These are the available factors of tourist attraction. In the sustainable development of the city, along with their promotion, protection is also necessary. The satisfaction of tourists is the question of credibility of any tourist place which decides the importance of its tourist place. Most of the tourists visit any tourist destination to experience its local supernatural attractions and mental relaxation. The level of satisfaction toward this first desire of the tourist was found to be high in the city of Udaipur in which the percentage of domestic tourists was high while foreign tourists were overwhelmed by the historical and cultural heritage here. Apart from this, the best experience for tourists of high-income group was due to the luxury ambience and royal services provided by the five and seven star hotels available in the city. Also, due to good services provided by other hotels in the city, a better level of satisfaction was found in each of the respective income groups. Tourists found this city more secure than other tourist cities and increased the sense of belonging in the hospitality here. Along with the above factors of extreme satisfaction, there are some factors in the city, toward which complete dissatisfaction was found among the tourists. Apart from the prestigious hotels of the city, dissatisfaction was found regarding the quality and hygiene of foods in other hotels and restaurants. Most of the tourists were found infuriate at the non-availability of local cuisines. At the same time, there was dissatisfaction among tourists regarding the behavior, knowledge and level of service of tour guides. Many tourists even questioned the role of tour guides at tourist places. Transport is the backbone of the sustainable economic development of any tourist city. According to the tourists, road infrastructure is inadequate as the city expands and number of vehicles increasing in a geometric progression. Poor traffic management system, disregard of traffic rules by vulnerable road users, careless vehicle drivers are factors responsible for dissatisfaction among the tourists. Inattentive and careless driving by auto-rickshaw drivers and exorbitant rates are the other reason for dissatisfaction toward public transport.

In addition to preserving and promoting tourism-related extreme satisfaction factors in the city, the administration and the local people will have to contribute for the encouragement of tourism development in the city by acting responsibly toward the dissatisfaction factors exposed in the tourism sector. There are so many locations available for ecotourism contenders for future ecotourism projects. There is still a lot of natural beauty left in and around Udaipur. What is required is a strong will from the administration, locals and tourists to save that beauty.

References

Alegre, J., & Garau, J. (2010). Tourist satisfaction and dissatisfaction. *Annals of Tourism Research, 37*(1), 52–73. https://doi.org/10.1016/j.annals.2009.07.001

Aliman, S. M. (2016). Tourist's satisfaction with a destination: An investigation on visitors to Langkawi Island. *4*, 1–20. Retrieved November 28, 2022, from https://www.researchgate.net/publication/303554285

Armario, E. M. (2008). *Tourist satisfaction: An analysis of its antecedents 367–382*. Retrieved December 19, 2022 from https://www.researchgate.net/publication/28236362

BBC Report. (2019). *The environmental impact of tourism*. British Broadcasting Corporation. Retrieved from https://www.bbc.com/news/environment:contentReference[oaicite:6]{index=6}

Chen, C.-F., & Chen, F.-S. (2010). Experience quality, perceived value, satisfaction, and behavioral intentions for heritage tourists. *Tourism Management, 31*(1), 29–35. https://doi.org/10.1016/j.tourman.2009.02.008

Customer Satisfaction. (n.d.). Customer satisfaction. Retrieved March 9, 2023, from http://www.customersatisfaction.com

Department of Tourism, Government of Rajasthan. (n.d.). Tourism in Rajasthan. Retrieved March 11, 2023, from http://www.tourism.rajasthan.gov.in

http://www.hondwhat.net/positive-negative-impacts-tourism/

Jingar, M. (2015). *A study of the factors influencing tourist satisfaction in Rajasthan.* Retrieved November 17, 2022, from www.ijcms.com

Ka Leong, C. (2016). *The impact of traffic congestions on tourist behaviour: Case study of Chiang Mai, Thailand.*

Khuong, M. N., & Ha, H. T. T. (2014). The influences of push and pull factors on the international leisure tourists' return intention to Ho Chi Minh City, Vietnam—A mediation analysis of destination satisfaction. *International Journal of Trade, Economics and Finance, 5*(6), 490–496.

Khuong, M. N., & Nguyen, T. D. (2017). The effects of destination image, perceived value, and service quality on tourist satisfaction and word-of-mouth: A study in Ho Chi Minh City, Vietnam. *Journal of Tourism and Hospitality Management, 5*(3), 276–290.

Kim, A. K., & Brown, G. (2012). Understanding the relationship between perceived travel experiences, overall satisfaction, and destination loyalty. *Anatolia International Journal of Tourism and Hospitality Research, 23*(3), 328–347.

Le, C. C., & Dong, D. X. (2017). Factors affecting European tourists' satisfaction in Nha Trang city: Perceptions of destination quality. *International Journal of Tourism Cities, 3*(1). https://doi.org/10.1108/ijtc-04-2017-0022

Leili Tapak, H. A. (2019). Assessment of factors affecting tourism satisfaction using K-nearest neighbourhood and random forest models. *BMC Research Notes, 1–5*, 4799. https://doi.org/10.1186/s13104-019-4799-6

Liyun Zeng, R. Y. (2021). Tourist satisfaction, willingness to revisit and recommend, and mountain kangyang tourism spots sustainability: A structural equation modelling approach. *Sustainability, 13*, 10620

Mdusman, C. (2016). The positive and negative effects of tourism on the social-cultural environment of a country. *The Journal of Tourism Studies, 28*(2), 123–140.

Ministry of Tourism, Government of India. (n.d.). Tourism in India. Retrieved March 13, 2023, from http://www.tourism.gov.in

National Trust. (n.d.). Heritage tourism. Retrieved March 11, 2023, from http://www.nationaltrust.org/heritage_tourism

Pal, S., & Rawal, Y. S. (2020). Perception and attitude of the host community on socio-economic development through tourism: A case study on Udaipur, Rajasthan. *International Journal of Advanced Science and Technology, 29*(3), 9202–9217.

Pavlic, D. P. (2015). *Tourists satisfaction as an important tool for the increasing tourism destination competitiveness in the globalization conditions: A case of Dubrovnik-Neretva.* Retreived November 19, 2022, from http://wwwresearchgate.net/publication/263750038

Pizam, A., & Neumann, Y. (1978). Dimensions of tourist satisfaction with a destination area. *Annals of Tourism Research, 5*, 314–322.

Rahman, M. S., Hasan, H., Gani, O., Abdel, F., & Anwar, A. (2017). Edu-Tourist perceived service quality and perception: The mediating role of satisfaction from foreign student's perspectives. *Tourism Review, 72*(2), 156–170.

Shahrivar, R. B. (2012). Factors that influence tourist satisfaction. *Journal of Travel and Tourism Research,* 1–20. Retreived November 27, 2022 from http://www.researchgate.net

Suanmali, S. (2014). *Factors affecting tourist satisfaction: An empirical study in the Northern part of Thailand, published by EDP Sciences*, 1–9. Retrieved December 12, 2022 from https://www.researchgate.net/publication/307710162

Times of India. (2022). G20 Sherpa meeting in Udaipur: Climate finance, tourism, culture on agenda. Times of India. Retrieved from https://timesofindia.indiatimes.com/india/g20-sherpa-meeting-in-udaipur-climate-finance-tourism-culture-onagenda/articleshow/96029157.cms

Trinanda, O., & Sari, A. Y. (2020). Exploring tourist satisfactions in eco-tourism: Antecedents and consequences. In *Proceedings of the 4th Padang International Conference on Education, Economics, Business and Accounting (PICEEBA-2 2019)* (pp. 988–993). Atlantis Press. https://doi.org/10.2991/aebmr.k.200305.168

Um, S., Chon, K., & Ro, Y. (2006). Antecedents of revisit intension. *Annals of Tourism Research, 33*(4), 1141–1158.

United Nations. (2015). Sustainable Development Goals: 17 Goals to Transform Our World. United Nations. Retrieved from https://www.un.org/sustainabledevelopment/sustainable-development-goals/:contentReference[oaicite:0]{index=0}

United Nations Environment Programme (UNEP). (2019). *Global environment outlook—GEO-6: Healthy planet, healthy people*. Cambridge University Press. Retrieved from https://www.cambridge.org/core/books/global-environment-outlook-geo6-healthyplanet-healthy-people/8FE2F127F310561C679B620F1D2EDBA6

Wang, T. L., Phuong, K. T., & Tran, V. T. (2017). Destination perceived quality, tourist satisfaction and word-of-mouth. *Tourism Review, 72*(4), 392–410.

Yuza, O. T. (2019). Exploring tourist in eco-tourism: antecedents and consequences. In *Proceedings of the 4th Padang International Conference on Education, Economics, Business and Accounting* (pp. 988–993).

Chapter 11
Delineation of Suitable Land for Agricultural Sustainability of Veniar River Basin Based on Geographic Information System

Pawan Kumar, Manish Kumar, Shikha Sharma, Krishna Kumar, Preeti Sachar, and A. K. Mishra

Abstract Suitability analysis for agricultural land use planning is a prerequisite to increase crop production for an ever-growing population. Agricultural land use planning by using GIS helps to decide on the best location to grow the crops. By selecting the best location for agriculture to increase crop production. It also helps the farmers without losing or damaging the crops. Here the analysis has taken GIS for the kind of parameters which is essentially needful for taking the specific land to take agricultural practices. The setting of parameters by using two technologies is GIS and remote sensing. The thematic data sets are the parameters: land use, soil type, slope, vegetation, and hydrology. Integration of these data sets is converted into the raster and classified into the subcategories in the selected parameters. Analytic hierarchy process (AHP) method and weights evaluate the criteria by using GIS raster overlaying technics to derive the suitable site. GIS overlay in spatial analysis operation was also used to combine the different map layers for the most suitable lands for agricultural development. In this seminar research paper, a hierarchy model was developed to identify the suitable area for agricultural land use planning using soils, slope, hydrology, and geomorphology of the area to support decisions concerning

P. Kumar (✉) · A. K. Mishra
Department of Geography, Delhi School of Economics, University of Delhi, Delhi, India
e-mail: pawank.dse@gmail.com

M. Kumar
Department of Political Science, Swami Shraddhanand College, University of Delhi, Delhi, India

S. Sharma
Departments of Political Science, Jamia Milia Islamia, Delhi, India

K. Kumar
Discipline of Geography, School of Sciences, IGNOU, New Delhi, India

P. Sachar
Department of Geography, Swami Shraddhanand College, University of Delhi, Delhi, India

© The Author(s), under exclusive license to Springer Nature Switzerland AG 2024
B. V. Raj Singh and A. K. Batar (eds.), *Sustainable Local Development for Environmental and Social Sustainability*, Human-Environment Interactions 11,
https://doi.org/10.1007/978-3-031-67303-0_11

the implementation of agricultural planning. GIS was used based on a set of criteria derived from the spatial and environmental aspects.

Keywords GIS · Remote sensing · Overlay analysis · Spatial decision support system

11.1 Introduction

Global Processing Systems (GPS), geographic information systems (GIS), and remote sensing (RS) need no letter of credence for their importance to be used as a valuable tool in land-use planning and related activities. GIS has been coupled to a variety of models (e.g., crop productivity, hydrology, and water quality simulations) and is an important component of spatial decision support system (SDSS) and land resources information system (LRIS) (Ebaid, 2008). GIS along with multi-criteria decision techniques on a wide range of areas of interest from the technical aspect of integrating GIS with multi-criteria evolution, Land suitability mapping for industry, peri-urban agriculture, bee-keeping, etc.) (Khoi, 2010). To derive the land suitability ratings by matching crop requirements data with land quality data, it establishes usually physical resource databases: soil, climate, land suitability, and productivity) (Doucet, 2011). Land suitability evaluation first by matching the biophysical factors affects the crop growth (climate, terrain, and soils) to the requirements of the crop. Suitability evaluation does not identify a single farm-use best on each land unit. Suitable classes for different uses cannot be compared in a routine automatic manner. The physically and economically viable alternatives have to be considered for planning decisions. The importance of land capability studies lies in their application for agricultural land-use planning. The planning. The planning could be done broadly for large areas using classes and sub-classes for smaller categories (Bunruamkaew et al., 2012). This study is to explain the process of developing a GIS application to provide a system for supporting location decisions concerning the implementation of agricultural planning. The model was developed to assess the suitability of the area for farmland use and to facilitate decision-making for sustainable agricultural production, considering the variables of slope, soil, and water availability. Sustainable agricultural production refers to a balance between land resources and their production capabilities. This helps the growing population to maintain food security, thereby emphasizing the need for a study on sustainable agriculture. Land use planning is a prerequisite for proposing the way out for sustainable agriculture. It is essentially based on land suitability analysis which is an assessment of an area on its appropriateness for a particular land use.

11.2 Aim and Objectives

The paper aims to look into the physical factors and delineate suitable areas for agricultural land use. Toward this goal, the study plans refer to the locational attributes needed for farmlands, attempt a suitability analysis based on analytic hierarchical process, and assess and rank the potentials of farmlands as good, moderate, low, and not suitable.

11.3 Method and Model

The suitability derived from using data sets taken from the satellite image of LANDSAT 8 OLI 2013 for extract the layers such us built-up lands, soil capability, and land use visual interpretation has been done, and then the hydrological layer is taken from ASTERGDEM which gives 30 and 15-m horizontal resolution to get streams and drainage density. After the collected data sets, the criteria are drawn for the associated suitable index for each layer.

The method adopted here to assess and locate the suitability of the land for agriculture is sketched in the flowchart above and that was differentiated into four segments. (1) The first one is the raw data used to derive the data extracted source where the data could extract. (2) The second segment is including the data converted into the raster and that data were classified based on the layer's suitability. (3) Third segment deals with the resultant data obtained then reclassified five classes with equal interval method for all layers then assigning weight percentages given based on their importance for suitability with the help of ArcGIS s/w spatial analysis tools. (4) The result was get from the reclassified layer with the raster cell-based overlay function (Fig. 11.1).

11.4 Study Area

Thirumanimuthar sub-basin is located which lies between 10.58 N and 11.48 N and 77.53 E and 78.28 E, with a drainage area of about 2438 km^2. Thirumanimuthar River originates at Manjavadi in Salem District and confluences at the river Kaveri at Paramathi in Namakkal District. The annual rainfall of this sub-basin is about 160 cm (Fig. 11.2).

The groundwater of the Thirumanimuthar region is a unique example of the impact of weathering, ion exchange, and anthropogenic processes controlling groundwater chemistry. The chemical composition of groundwater in the study area is strongly influenced by rock-water interaction, dissolution, and deposition of silicate groups of minerals. Weathering of silicate minerals controls the major ion chemistry of calcium magnesium, sodium, and potassium.

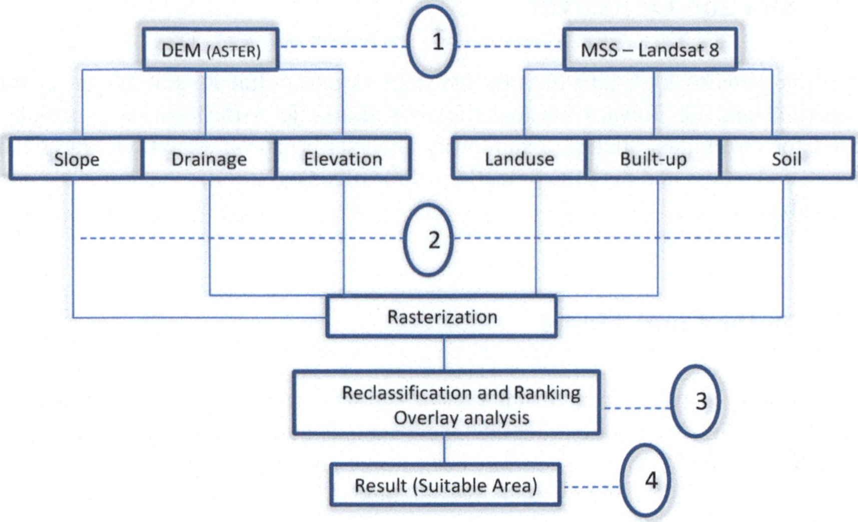

Fig. 11.1 Flow diagram of the methodology

The geology of the study area is composed of rock types represented by metasediments with charnockites, granitic gneiss, calc granulite syenite, and ultra-basic as major exposures. Hills, undulating plains, pleatues pediments, and bajadas are the mixture of geomorphic units of the study area. (Krishna, 2014) and the basic material of soil and granitic rock. Cl was higher in the post-monsoon season (1730 mg/L) due to leaching from upper soil layers due to industrial and domestic activities and dry climates) (Kanniyappan et al., 2020).

11.5 Factors of Land Suitability

There were six factors selected for basic needs to envelop the agricultural areas. Slope, elevation, drainage, soil, built upland, land, and use were used.

Slope

Slope defines the inclination of the topographic surface. A steep slope may promote erosion, and it may not promote farming activities. The slope layer was in raster format, so computation was performed on a cell-by-cell basis. The model function used for each criterion depends on the 'trend of performance' of the respective land attribute in giving a favorable condition for cropping (Malczewski, 2004). The flat surface is good for cropping because it holds the water and infiltrates into the soil.

Here the study classifies the topography of the study area into flat, moderate, and steep slopes to determine the suitability of agriculture; likewise the slope classified

Fig. 11.2 Location map of the study area. *Source* Kanniyappan et al. (2020)

based on an angle from 0 to 5 is too flat, and it can conceivable for practicing agriculture (Fig. 11.3). Slope angle of 5–9 is classified moderately suitable and more than 9–15 less suitable for cropping fields. The above 15 above is highly not suitable for agriculture. The analysis of slope angles of the study area has shown, and that 90% of the study area is fall under the suitable category of 0–5 angle, which is more suitable for agricultural use.

Drainage Density

It is the measure of how well or how poorly a watershed is drained by stream channels. Drainage density depends upon both climate and physical characteristics of the drainage basin. Rugged regions or those with high relief have a higher drainage density. High densities indicate greater runoff, less infiltration, and also greater flood

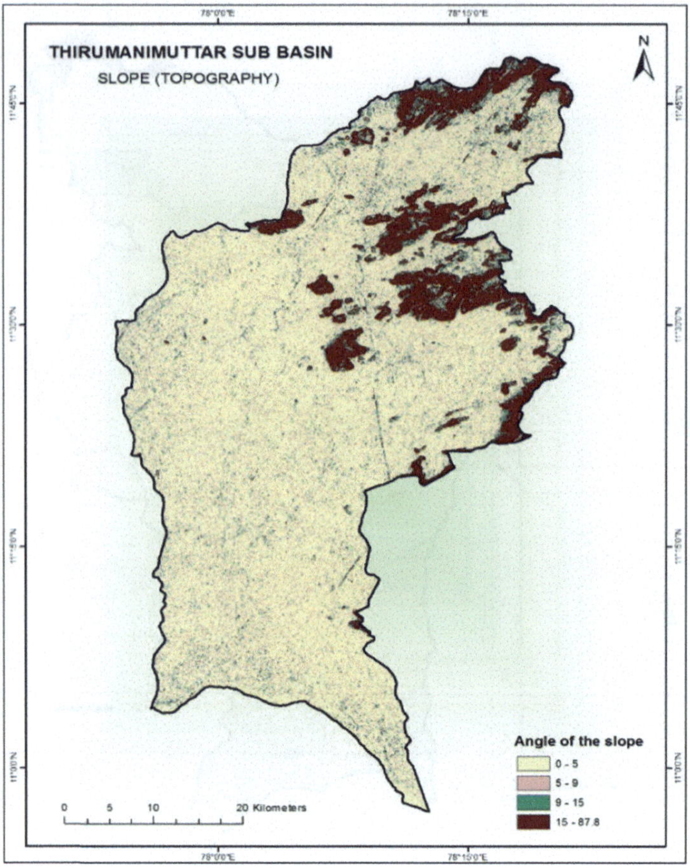

Fig. 11.3 Slope of Thirumanimuthar sub-basin

risk (Fig. 11.4). Drainage density is very high in the middle and lower regions of the basin and gradually decreases in the west and eastern regions. Almost 70 of the total area falls under the very low to moderate drainage density.

Built-Up Land Density

The built-up density refers to rural and urban regions in the study area the higher density of the built-up land is urban, and the lower density is referring the rural area. Here that is applied if the density is higher and is not suitable and the area has less density that is determined as suitability for agriculture. Here the kernel density is used to determine the built-up land density in the study area. Kernel density analysis employs the kernel density estimation (KDE) method to detect dense areas in a road network and uses the dense areas as built-up areas. The KDE also involves thresholds, i.e., bandwidth and percentage volume contours (Panda et al., 2010).

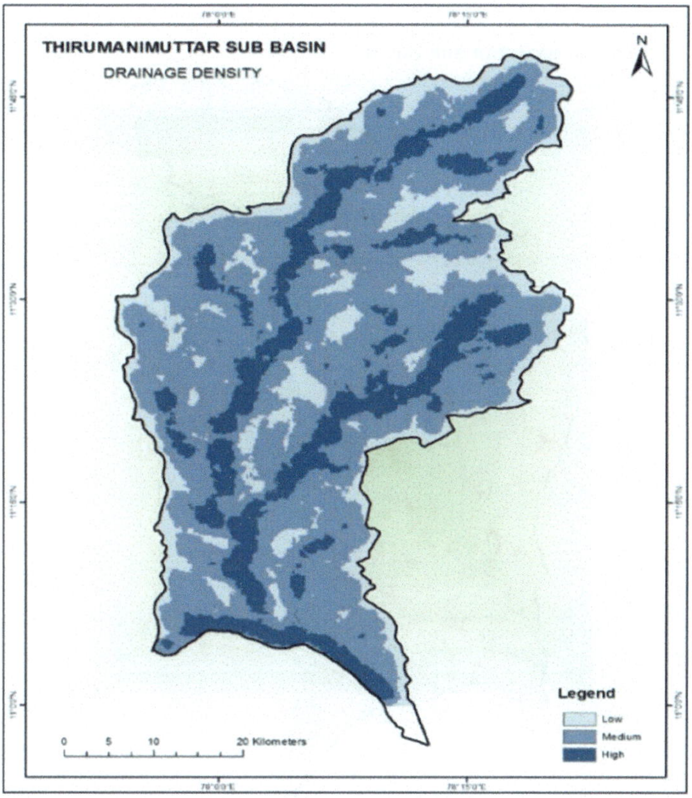

Fig. 11.4 Drainage density of Thirumanimuthar sub-basin category

Built-up density is very high in the upper region of the basin. The lower region is more rural, and so more land area is suitable for agriculture. Almost 60% of the total study area falls under the very low to moderate built-up density category (Fig. 11.5).

Soil

The soil is one of the main characteristics of agriculture. The healthy crops are mainly growing in the water presence area. Here potentials of soils are measured by two parameters: (1) Soil moisture stress index. Soil moisture is one of the most important factors that define the presence of earthworms in the soil. One of the consequences of an increased earthworm population is the formation of channels and pores. The consequence of increased organic matter content (and microbial activity) is an increase in the earthworm population (FAO). Soil moisture is an essential variable governing the fluctuations of crop yield (Ronald Eastman et al., 2008). Crops are sensitive to soil moisture. The soil moisture deficit in the root zone during various stages of the growth cycle will have a profound impact on crop yield (Sallaku, 2008).

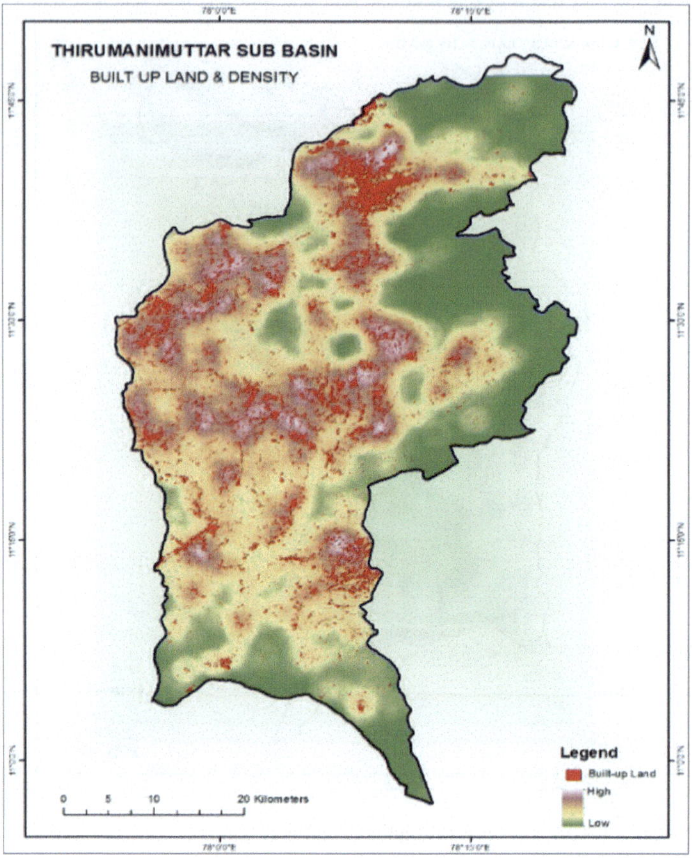

Fig. 11.5 Built-up area of Thirumanimuthar sub-basin

$$\text{MSI} = \frac{\text{Mid IR}}{\text{NIR}}$$

Jackson this formula is used to develop a moisture stress index to proceed with one of the parameters of agriculture suitability (Fig. 11.6).

Land Use

Agricultural production has caused a greater environmental change to the biosphere than any other land use. Landuse use classes were used to derive suitable areas for agriculture. That has been done by assigning priority to each land-use class. For example, high importance is given to the fallow, current fallow, and cropland and less importance to forests, waterbodies, and pastureland (Fig.11.7).

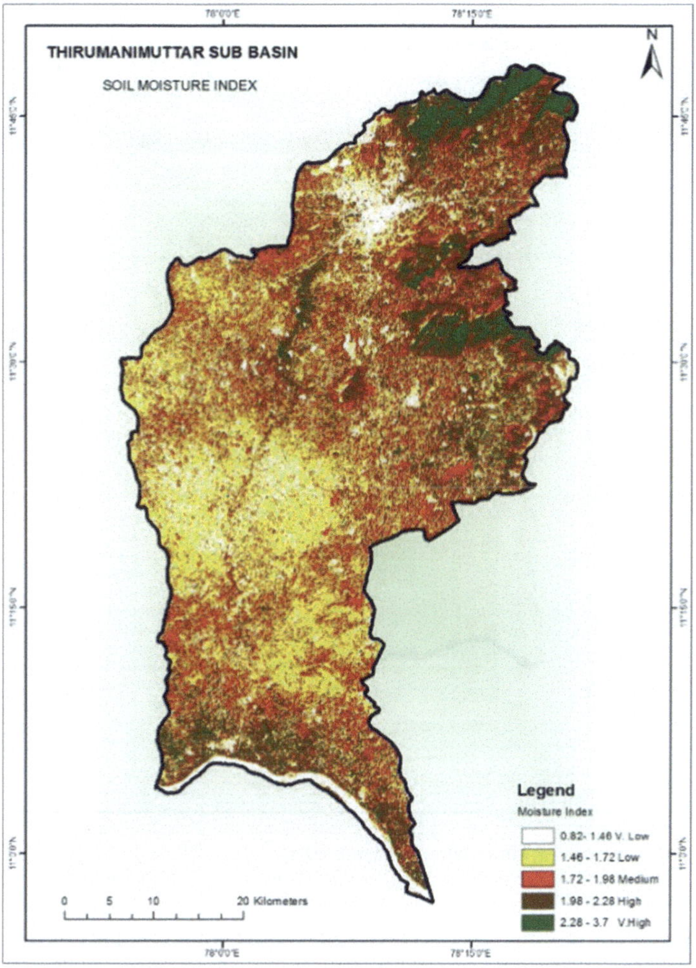

Fig. 11.6 Soil moisture of Thirumanimuthar sub-basin

11.6 Land Suitability Map Build with GIS

In this study, remote sensing and GIS techniques were used to identify agriculturally land suitability areas for Thirumanimuthar. The study was carried out to delineate the spatial distribution of agricultural land-use suitability areas developed with an evaluation of parameters like soil, slope, drainage density, and land use. These parameters have been extracted from the ASTERGDEM and LANDSAT 8 satellite images. Then all the resultant data had evaluated based on agriculture suitability. Then with the help of a raster-based overlay method is adopted to achieve a suitable result. Then it has been calculated based on suitable classes, and it was successfully visualized. The suitability level was quantitatively represented in the following table and graph.

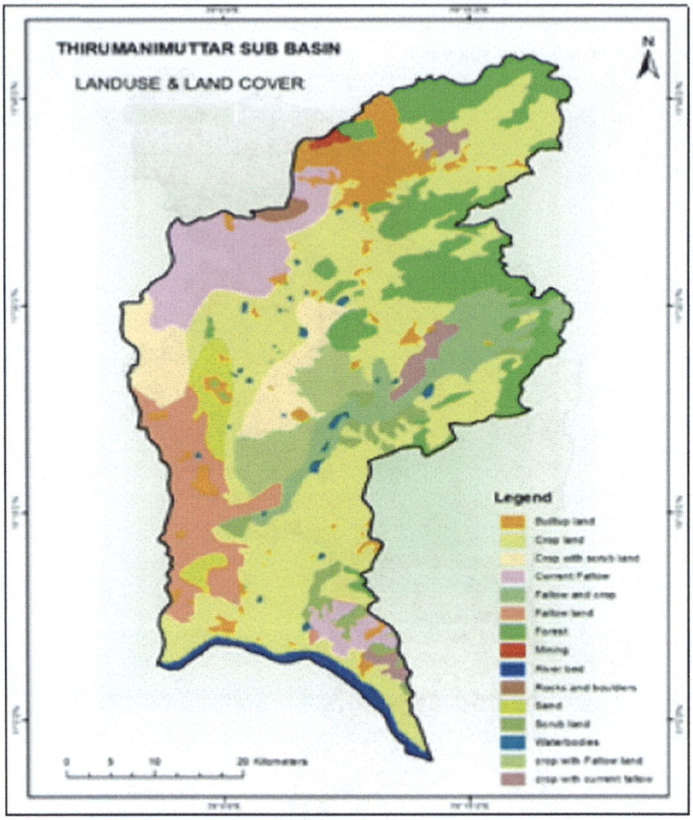

Fig. 11.7 Land use/cover of Thirumanimuthar sub-basin

It is observed that the suitable areas are closer to the southeast part of the region in the sub-basin (Fig. 11.8). The suitability grid of this scenario was reclassified to five classes. Not suitable areas represent (15.5%) of the study area, and the less suitable area is (11.2%); this land comes under the forest, and protected areas in the study area were not applicable for agricultural development. Moderate suitability is (20.67%).

Terminology	Area in hectares	%
Not suitable	37,512.3	15.54652
Less suitable	27,174.9	11.26231
Moderate	49,823.7	20.64883
More suitable	76,044	31.51551
Very much suitable	50,735.8	21.02684

These regions would be the land that can practice for agriculture but it will not be used because in this category slope is not stable; this will be suitable for plantations

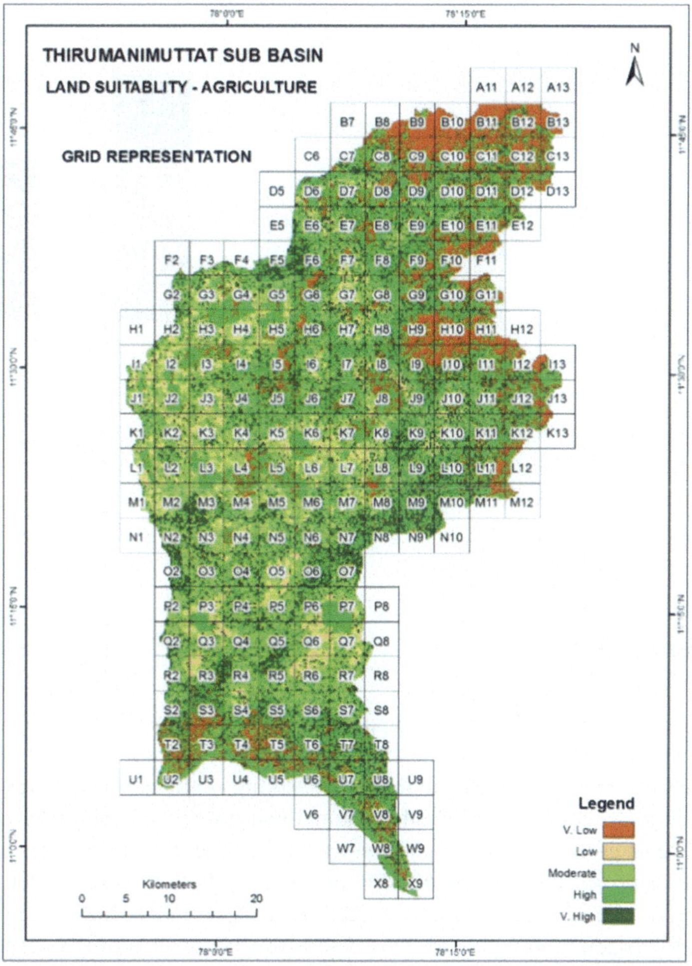

Fig. 11.8 Land suitability area of Thirumanimuthar sub-basin

which are crops growing with less water. More suitable (31.5%), and very high suitable (21.8%) these areas which have the all selected factors contribute.

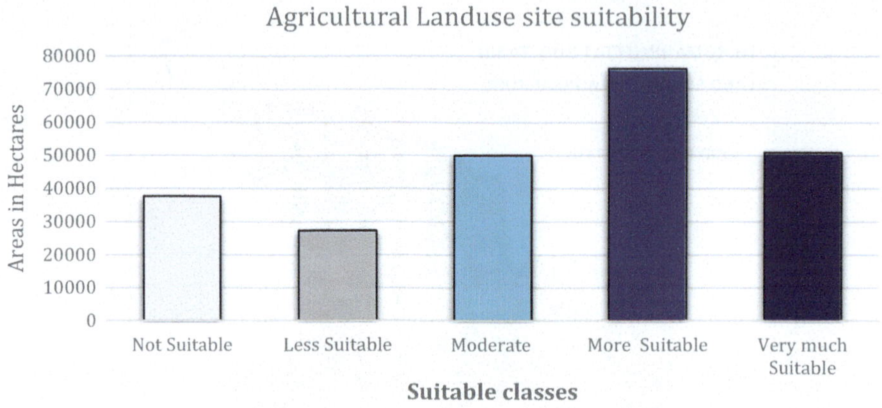

It is highly suitable for agriculture crop managements like soil moisture, slope, drainage, and open space. This land is too potential land for agriculture. This will help farmers more productivity and agriculture planners and policymakers.

11.7 Conclusion

Global climate change makes some impact on different types of crops and therefore affects production and productivity and ultimately the world's food supply. Substantial study has undergone to work out that how farming is affected in different parts of the regions, and by how much; whether the net result may be harmful or beneficial to which type of crop. However, major uncertainties limit the correctness of current models and or projections. Some of them are relating to the degree of temperature increase and its geographic distribution and other may pertain to links changes likely to occur in the precipitation patterns that determine the water supply to crops and to the evaporative demand imposed on crops by the warmer climate. Some areas could experience longer, more intense droughts, resulting from higher summer temperatures and reduced precipitation makes difficult to crop production. In Indian condition, climate change and its variability from place to place are placing challenges and ultimately impacting Indian agriculture. Reviewed from the many article and books mentioned site suitability analysis performed within GIS and remote sensing environment was successful. It will be useful for many applications such as habitats, forestry, real estate, urban environments, and so on. The GIS-Multi-criteria approach which however does not affect spatial land use scenarios and is sustainable for future land use planning is optimizing the environmental benefits. The land-use suitability analysis using GIS-based multi-criteria analysis showed high potential for use in land use planning.

References

Bunruamkaew, K., & Murayama, Y. (2012). Land use and natural resources planning for sustainable ecotourism using GIS in Surat Thani, Thailand. *Journal of Sustainability, 4*, 412–429.

Doucet Beer, E. (2011). *Modeling alternative agricultural scenarios using RUSLE and GIS to determine erosion risk in the Chippewa River Watershed, Minnesota. Michigan.* Natural Resources and Environment: The University of Michigan

Ebaid, H. M. (2008). Assessment of soil suitability for agriculture using GIS spatial analyst model technique. In *The 3rd international conference on water resources and arid environments*. Center Giza

Kanniyappan, S. P., et al. (2020). Identification of groundwater recharge zones in Thirumanimuthuar, sub-basin in parts of Salem and Namakkal District, Tamilnadu, India. *International Journal of Research in Engineering and Science, 08*(5), 25–36.

Khoi, D. D. (2010). Delineation of suitable cropland areas using a GIS-based multi-criteria evaluation approach in the Tam Dao National Park Region. *Vietnam. Journal of Sustainability, 2*, 2024–2043.

Krishna, G. (2014). Agricultural land suitability analysis of a River Basin Area. *International Journal of Geospatial Engineering and Technology, 12*, 37–42.

Malczewski, J. (2004). *GIS-based land-use suitability analysis: A critical overview*. Elsevier.

Nyeko, M. (2012). GIS and multi-criteria decision analysis for land use resource planning. *Journal of Geographic Information System, 04*, 341–348.

Panda, S. S., & Hoogenboom, G. (2010). Remote sensing and geospatial technological applications for site-specific management of fruit and nut crops: A review. *Journal of Remote Sensing, 2*, 1973–1997.

Ronald Eastman, J., & Toledano, J., et al. (2008). Participatory multi-objective decision-making in GIS. In *The Clark labs for cartographic technology and geographic analysis*. Clark University.

Sallaku, F. (2008). *The agricultural land suitability assessment as the main factor for land consolidation activities in Albania*. Regional Agricultural Extension Centre, Ministry of Agricultural and Food.

SPRINGER NATURE

GPSR Compliance

The European Union's (EU) General Product Safety Regulation (GPSR) is a set of rules that requires consumer products to be safe and our obligations to ensure this.

If you have any concerns about our products, you can contact us on ProductSafety@springernature.com

In case Publisher is established outside the EU, the EU authorized representative is:

Springer Nature Customer Service Center GmbH
Europaplatz 3
69115 Heidelberg, Germany

The manufacturer's authorised representative in the EU is Springer Nature Customer Service Centre GmbH, Europaplatz 3, 69115 Heidelberg, Germany. If you have any concerns regarding our products, please contact ProductSafety@springernature.com

Printed and bound by CPI Group (UK) Ltd, Croydon, CR0 4YY
01/10/2026
02225961-0002